Ecological Studies

Analysis and Synthesis

Edited by

W. D. Billings, Durham (USA) F. Golley, Athens (USA)

O. L. Lange, Würzburg (FRG) J. S. Olson, Oak Ridge (USA)

Volume 31

W. Tranquillini

Physiological Ecology
of the Alpine Timberline

Tree Existence at High Altitudes
with Special Reference to the European Alps

With 67 Figures

Springer-Verlag Berlin Heidelberg New York 1979

Prof. Dr. Walter Tranquillini
Forstliche Bundesversuchsanstalt
Außenstelle für subalpine Waldforschung
Rennweg 1, Hofburg, A-6020 Innsbruck

Translated by:
Dr. Udo Benecke
Forest Research Institute
Christchurch, New Zealand

For explanation of the cover motive see legend to Fig. 61 (p. 110).

ISBN 3-540-09065-7 Springer-Verlag Berlin Heidelberg New York
ISBN 0-387-09065-7 Springer-Verlag New York Heidelberg Berlin

Library of Congress Cataloging in Publication Data. Tranquillini, Walter, 1924–. Physiological ecology of the alpine timberline. (Ecological studies; v. 31) Bibliography: p. Includes index. 1. Timberline. 2. Trees—Physiology. 3. Forest ecology. 4. Alpine flora. 5. Timberline—Alps. 6. Trees—Alps—Physiology. 7. Forest ecology—Alps. 8. Alpine flora—Alps. I. Title. II. Series. QK938.F6T69. 581.5'264. 78-25603.

Typesetting, printing, and binding: Brühlsche Universitätsdruckerei, Lahn-Gießen.
2131/3130—543210

In memory of Prof. A. Pisek,
a pioneer of alpine ecology

Preface

In the European Alps the importance of forests as protection against avalanches and soil erosion is becoming ever clearer with the continuing increase in population and development of tourism. The protective potential of the mountain forests can currently only be partially realised because a considerable proportion of high-altitude stands has been destroyed in historical times by man's extensive clearing of the forests. The forests still remaining are of limited effectiveness, due to inadequate density of trees and over-maturity. Considerable efforts, however, are now being made in the Alps and other mountains of the globe to increase the high-altitude forested area through reforestation, to raise depressed timberlines, and to restore remaining protection forests using suitable silvicultural methods to their full protective value.

This momentous task, if it is to be successful, must be planned on a sound foundation. An important prerequisite is the assembly of scientific facts concerning the physical environment in the protection forest zone of mountains, and the course of various life processes of tree species occurring there. Since the introduction of practical field techniques it has been possible to investigate successfully the reaction of trees at various altitudes to recorded factors, and the extent to which they are adapted to the measured situations. Such ecophysiological studies enable us to recognize the site requirements for individual tree species, and the reasons for the limits of their natural distribution.

During many years of research at alpine timberline in Tyrol, Austria, it has been possible for me to publish numerous articles on the ecophysiology of timberline trees. It is to one of the pioneers of experimental ecology, Professor A. Pisek, that I owe my introduction to this field of research and this book is dedicated with gratitude in his honour.

Colleagues at the Institute for Subalpine Forest Research in Innsbruck effectively supported my work and have brought together a mass of information on the ecology of timberline which has been drawn on in preparing this book. In the last 10 years students of the University of Innsbruck acted as valuable assistants and their dissertations stand as important contributions. It now appears useful to draw the numerous and diverse published research results together into a single volume.

In Europe I became familiar with the timberlines of the major mountain ranges besides the Alps and visited the main timberline investigations currently in progress. Discussions of the upper forest limits on mountains outside Europe are mainly based on an evaluation of the literature.

In treating the subject matter, particular emphasis has been placed on describing as completely as possible the interrelationship between site factors and the most important life processes, in order to elucidate the significance of such factors for the existence of trees at their upper limit. This approach is possible since site factors at timberline, as at other distinct vegetation limits, gain in significance, and clearly become limiting factors for vital plant processes. In contrast to this the importance of competition declines.

Although there are still gaps in our knowledge, an attempt has been made in the Synopsis (Chap. 6) to develop from numerous individual results a general picture of the life of trees at the limits of existence on mountains. The complex framework of interrelationships responsible for the occurrence of timberlines may thus be seen with greater clarity.

Much in the concepts presented remains unproven and hypothetical. Inevitably, the picture evolved here will not be valid everywhere, and other mechanisms may exist for formation of timberlines. The great diversity of environments and tree species on the earth's mountain ranges has been far too sparsely investigated for one to conclude otherwise. For the European Alps, however, as probably for other temperate mountains, this volume will hopefully form a valid contribution to the scientific foundation upon which reforestation and management of protection forests must be based.

Acknowledgements. For reading the manuscript and making many valuable suggestions towards its improvement I wish to thank Dr. W. M. Havranek.

My special thanks must go to Director Dipl. Ing. J. Egger, whose generous support of my work over the years significantly contributed to the completion of this volume.

To the publishers I also wish to express my appreciation for their co-operation and willingness to consider all my requests.

Innsbruck, January 1979   W. Tranquillini

Contents

General Features of the Upper Timberline

In many parts of the earth there are natural barriers to the spread of forests. In subpolar regions it is cold temperatures, in the subtropics it can be drought, and in temperate zones it may be extreme soil conditions, such as salinity and water logging, or on sea coasts and on the tops of isolated hills it can be fierce winds which halt the forest (Ellenberg, 1966).

Visually the most striking vegetation boundary is the upper limit of forests wherever mountains rise to a sufficient height (Hermes, 1955). The form this upper timberline takes is very varied. The forest may reach its upper limit as a closed stand, and cease abruptly as a sharp line against a treeless alpine zone. It may also however, gradually dissolve from a dense high stand to isolated trees and finally to stunted individuals, thus presenting a broad transition zone. In this case one must differentiate between a forest-, tree-, and krummholz-limit (Tranquillini, 1976). This transition zone above the closed stand is normally designated the "kampfzone" where trees struggle for their existence, as witnessed by varying degrees of damage leading to crippled dwarf tree forms at the uppermost limits. In the English literature this transition zone above the forest or timberline ecotone, including both the tree- and krummholz-limits, is often referred to as the krummholz zone (cf. Fig. 61).

Opinions still differ as to which of the two types of timberline, i.e., with or without a transition zone, represents the natural situation uninfluenced by man (Plesník, 1971). Of the two opposing theories (Schröter, 1926, p.40), one suggests that the gradual opening of the stand results from a worsening of the factors for growth and survival with increase in altitude. Isolated trees receive more light and heat, ensuring greater productivity than is possible in the closed stand. They are, therefore, capable of existing at altitudes above the stand (Frankhauser, 1901). A consequence of trees standing in isolation is winter damage, which higher up the slope leads to increasingly stunted growth and the krummholz limit. A broad transition zone would accordingly be a natural and primary condition.

The alternative theory states that wherever a single tree occurs a closed forest can also exist (Scharfetter, 1918). Thus wherever soil and topography allow, the forest will continue as a closed stand up to its final upper limit, forming a sharp boundary with the alpine grassland (Ellenberg, 1966; Schiechtl, 1966). A fully stocked forest generates its own even and relatively favourable internal climate for the existence of trees. This ameliorated climate contrasts with the climatic extremes of the open grassland. Young trees establishing outside a protective forest canopy above the forest limit are rapidly eliminated and the primary transition zone remains very narrow (Fig. 1).

Fig. 1. Natural timberline, barely influenced by man, in the Radurscheltal (Tyrol, Austria) of almost pure *Pinus cembra* at 2100–2200 m a.s.l. The dense stands show little sign of opening up towards their upper limit, and thus form a sharp boundary. The form of land tenure and difficulties in timber extraction have resulted in very limited utilisation of the forests in this region

Adherents to this second theory maintain that a gradual opening of the forest canopy and a broad timberline ecotone are secondary phenomena induced mainly by man's past destructive influence (fire, grazing, and felling). Present-day timberlines are thus clearly not as high as the climate potentially allows (Schiechtl, 1966). If such an anthropogenic timberline is left undisturbed, then the forest soon begins to reinvade upwards (Mayer, 1976), but this is accompanied by severe damage to young trees due to the climatic extremes existing on such deforested slopes (Fig. 2). Eventually this secondary transition zone would transform into a closed forest and the natural abrupt upper boundary will be restored.

These differing opinions from the European Alps take on a new aspect when viewed in the light of timberlines in other parts of the globe (Wardle, 1974). It becomes immediately clear that the physiognomic diversity of this vegetation boundary is also dependent on the greatly differing site requirements of the different tree species. Not all species show the same reaction to similar limiting climatic conditions. *Pinus flexilis* and *Pinus aristata* in the Rocky Mountains of Colorado, for instance, develop into erect trees amongst prostrate krummholz of *Picea engelmanni* and *Abies lasiocarpa* because they appear to withstand winter desiccation (Wardle, 1965). The differing ecological requirements of tree species can also lead to the formation of a double timberline. In the Great Basin of N-W America *Pinus monophylla* forms timberline

Fig. 2. Anthropogenic timberline in the Viggartal (Tyrol, Austria) of *Picea abies, Larix decidua* and *Pinus cembra* at 1900 m a.s.l. The high altitude stands were most probably felled during the middle ages to supply timber for mining, and this resulted in the destruction of the cembran pine belt. In subsequent centuries the forest has spread back up the slopes, forming a wide transition zone (kampfzone) above the forest. Tree limit lies at 2100 m and the krummholz limit attains 2200 m a.s.l. (Stern, 1966)

with a transition zone at 2250–2400 m a.s.l. Well above this altitude, stands of *Pinus flexilis* and *Pinus aristata* extend to their upper limit at 3450 m (Wardle, 1965; Andre et al., 1965). Generally, shade-tolerant species tend to form sharp timberlines, while light-demanding species tend to produce open stands near timberline with a wide transition zone (Walter, 1968, p. 478).

One must also remember that environmental conditions are not uniform at the upper limits of the earth's forests. This too can contribute to diversity in physiognomic form of these upper stands. In parts of the Rocky Mountains where the forest canopy has barely been touched by man, one can frequently find an extensive transition zone above the forest (Sharpe, 1968). Walter (1971/1972) also describes for eastern slopes of the Pacific Mountains in North America a totally undisturbed region where the natural forest does not form a sharp boundary, but instead a mosaic of tree groups decreasing in stature with increasing altitude. The reason for such a pattern appears to lie in the uneven distribution of snow. Trees can only survive on raised ground with early snow-melt, and in depressions the snow-free period is too short for tree growth.

The great influence of climate on the physiognomy of timberline becomes particularly evident on tropical mountains where seasonal climatic fluctuations are largely absent, but diurnal patterns are very marked. Climatic features, such as lack of a winter season and absence of snow cover, strongly modify the influence that limiting heat exerts on the forests, and lead to great diversity in the character of tropical timberlines. In some cases trees become ever shorter with increasing altitude, and the vegetation transforms very gradually via scrub of diminishing size to tall grass. Here, however, an exact definition of timberline becomes very difficult (Troll, 1966, 1973).

Reasons for Occurrence of Timberlines
and Their Experimental Investigation

Though reasons for upper limits of different tree species can vary, including local variations in contributory factors, timberline is ultimately dependent everywhere on the increasingly unfavourable heat balance with rising elevation above sea level. In regions with seasonal climates, increasing cold lengthens the frost period and shortens the frost-free period available for plant production. In addition, whatever adaptation to cold a tree may have, its growth and development is restricted during the vegetative period because heat no longer suffices for completion of the growth cycle. The same applies in principle to trees at the polar treeline. The long winter blockade to life processes is absent in the tropics, but is substituted by inhibition of growth through nightly frosts (Troll, 1966).

There are a number of basic observations demonstrating that heat balance of sites greatly influences the distributional limits of forest. Thus timberline in a valley lies at higher altitude on the slope with sunny aspect than on the one with shady aspect (Schröter, 1926, p. 29; Köstler and Mayer, 1970). Towards the valley head timberline drops ("valley-head phenomenon", Scharfetter, 1938) due to cold air currents converging from above, particularly where valleys are glaciated (Friedel, 1967). This effect also contributes to the warm mid-slope zone tapering towards the valley head (Böhm, 1969; Holtmeier, 1974). Air pockets chilled by the exit of cold air from ice holes induces local occurrence of vegetation normally found at higher levels, and can cause spruce to be unthrifty (Furrer, 1966). Ponding of cold air in basins can lead to inversion of the altitudinal vegetation zonation with a cold-induced lower timberline. Well-known examples of this are the Gstettner meadow depression (sink hole) of the East Alps (Gams, 1935) and the meadow hollows in the Ternovaner Forest near Görz (Tschermak, 1950, p. 12). Other examples of inverted timberlines are described by Wardle (1971).

There is a very noticeable climb of the upper forest limit from the outer chains of a mountain massif to the climatically more continental central ranges with their relatively warmer summers (Brockmann-Jerosch, 1919). There is no better demonstration of this overriding influence of temperature on altitude of timberline than the obvious rise in timberline from the polar regions to the subtropics (Daubenmire, 1954; Hermes, 1955; Ellenberg, 1966).

Comparisons in older literature between timberline location and certain isotherms had already indicated that temperature during the growing season plays the key role in setting the forest limits. The 10° C July-isotherm agrees relatively well with the course

Fig. 3. Ecological field station "timberline" of the Federal Forest Research Institute, Austria (Division of Subalpine Forestry) in the Gurglertal, Tyrol. The instrumented base station and main trial sites are located immediately above the timberline of *Pinus cembra* at 2070 m a.s.l. Additional instrumented sites were established in the closed forest below and in the timberline ecotone (kampfzone) above this base station. Climatological, soil, silvicultural, and plant physiological studies began in 1953 and continued intensively until 1972

of alpine and polar timberlines. From this one can deduce a rough rule that tree growth is possible wherever the mean temperature for the month of July exceeds 10° C. An even closer agreement is obtained between timberline and a mean daily maximum temperature of 11.1° C during the growing season, and thus mid-day temperatures in summer have similar values at all timberlines (Marek, 1910).

Such climatic iso-lines rarely agree exactly with the course of timberline and do not yield a satisfactory explanation of causes for the timberline phenomenon for the following reasons:

1) Climatic data used for such comparisons largely originate from meteorological stations usually remote from timberline, so that reliance has to be placed on inaccurate estimations of altitudinal lapse rates. Air temperatures are measured in weather-screens and deviate from plant temperatures particularly in the mountains (Tranquillini and Turner, 1961). Calculations made by Davitaja and Melnik (1962) of annual temperature summations of daily means above 10° C are instructive in this respect. Polar timberline was correlated with a mean air temperature sum of 600–700° C whereas for alpine timberline the figure was 200–300° C. On a basis of leaf temperature, however, both polar and alpine timberlines lie where annual sums of leaf temperatures above 10° C attain about 800° C. Cooler mountain air temperatures at alpine timberline are thus compensated by marked increases in leaf temperature above

Fig. 4. The climate laboratory of the Federal Forest Research Institute, Austria, on Patscherkofel near Innsbruck at 1950 m a.s.l. Trees of *Pinus cembra* are in the foreground and form timberline immediately below the building. Cloud blankets the Stubaital in the background above which rise the glaciated mountains of the Central Alps. The laboratory houses facilities for investigating the physiological response of young woody plants to environmental parameters

ambient air temperature as a result of greater radiation intensities than at polar timberline.

2) The effect of temperature on plants is complex. It includes a direct influence on numerous physiological processes, e.g., germination, growth, photosynthesis, respiration, etc., which can have different minima, optima, and maxima. Temperature alters the physiological conditions of plants, and thereby their response to temperature. Very high and low temperatures ultimately lead to heat and frost damage, and in the extreme to death of plants. The relationships between temperatures and plant processes which determine survival and death are too numerous to be characterised by a simple mean value for temperature.

3) Temperature is not the sole factor acting on life of trees under limiting conditions on mountains. Hypotheses for causes of timberline have been variously reviewed (Marek, 1910; Schröter, 1926; Daubenmire, 1954; Hermes, 1955; Baig, 1972). Local variations in radiation, moisture, and wind factors modify the dominating influence of temperature, thus leading to deviations of timberline from isotherms.

These considerations lead to the conclusion that a causal analysis of upper forest limits is best attained by continuous observation of trees at their natural sites, and monitoring of their life processes in relation to the measured changes in environmental factors (Däniker, 1923). Only such ecophysiological studies at timberline allow identification of the plant's reaction to environmental extremes, their adaptation to any given set of conditions, and the limits beyond which adaptation is no longer possible because damage halts development somewhere from seed to mature tree. Investigations of this type were begun on timberline trees barely 50 years ago (see historical review) and to date have been confined to work in a limited number of research institutes (Karrasch, 1973). Inevitably many questions remain unresolved, and whole timberline regions, particularly in the tropics, have so far not been studied.

In more recent years, continuous recording stations have been established at timberlines, and modern mobile laboratories have been assembled in the service of ecological research. These facilities have enabled a concentrated interdisciplinary approach by specialists in various research fields to be made of the timberline problem as a whole. A considerable proportion of these investigations were completed at the Obergurgl timberline field station (Fig. 3) and in the climate laboratory at Patscherkofel near Innsbruck (Fig. 4), both described in the literature (Fromme, 1961; Tranquillini, 1965a). It has been the author's privilege to be actively involved with research at these two stations since their inception.

An attempt to evaluate critically results obtained, backed by personal experience, is thus felt justified. The aim is to present as comprehensive a picture of the life of trees at their upper limit as is possible. It is hoped thereby to produce a deeper insight into reasons for the occurrence of such a distinctive and important vegetational boundary.

Some Milestones in the History of Ecophysiological Research Concerning Alpine Timberline

First account using ecological research methods:

Däniker (1919–1920), Swiss Alps. Growth, morphology, and anatomy of timberline trees, analysis of physiological damage.

Pioneer work in experimental ecology:

Goldsmith and Smith (1926), Rocky Mountains, Colorado.
Michaelis (1931–1933), Kleines Walsertal, Germany.
Schmidt (1936), Black Forest, Germany.
Steiner (1935), Dürrenstein, Austria.
Pisek et al. (1935–1976), Patscherkofel near Innsbruck, Austria.

Team work at modern long-term research stations:

Federal Forest Research Institute, Austria, Subalpine Forestry Division:
Timberline Experimental Station near Obergurgl (established 1953) in the Ötztaler Alps, Tyrol.
Climate Laboratory on Patscherkofel near Innsbruck (established 1963).
Eidgenössische Anstalt für das forstliche Versuchswesen, Switzerland: Stillberg research station in Dischmatal near Davos (established 1959).

Modern Mobile Laboratories:

Mooney et al. (1963), White Mountains, California.
Slatyer and Morrow (1971–1972), Snowy River Valley, Kosciusko National Park, Australia.
Benecke and Havranek (1975–1976), Craigieburn Forest Park, Southern Alps, New Zealand.

Contributions to comparisons of different timberlines:

Holtmeier (1963 onwards), alpine timberline in Graubünden, Switzerland and timberlines in northern Scandinavia.
Wardle (1965 onwards), timberlines of the northern and southern hemispheres.

1. Natural Regeneration of Tree Stands at Timberline

At natural timberline there exists a dynamic equilibrium between advance of regeneration upwards in the timberline ecotone (kampfzone) and retreat of this regeneration due to existence of external conditions no longer conducive to survival. Natural regeneration thus plays an important role in maintaining the upper forest limit. Prerequisites to this end are the presence of seed-bearing trees, together with the occurrence of seed years resulting in adequate production and distribution of viable seed.

Berner (1959) emphasised the significance of seed production for retention of timberline. Vegetative growth can still be quite vigorous above the upper reproductive limit and ultimate tree age can even increase. Thus cembran pine (*Pinus cembra*) may attain an age of 250–300 years as an ornamental in parks, 350–400 years in mountain stands and 600–1200 years at extreme altitude. The upper limit to forest stands is, however, always determined in the end by restrictions to the regenerative capacity.

1.1 Seed Production. Frequency of Seed Years, Quantity and Quality of Seed

Formation of flowers, fruits, and seed demands large quantities of organic matter. Consequently in a season with strong reproductive growth, i.e., a mast year, the tree's reserves may be almost completely exhausted. How soon another prolific mast year can follow depends primarily on the time required for the tree to restock depleted reserves. Since photosynthetic performance decreases with increase in altitude (see Chap. 3), seed years become less frequent the higher the location of tree stands.

According to Tschermak (1950) spruce, *Picea abies*, produces cones at favourable sites every 3–5 years, at higher altitudes every 6–8 years, and at timberline only every 9–11 years. Seed-year interval at timberline for cembran pine, *Pinus cembra*, is recorded variously as 7–8 years (Oswald, 1963) and 3–10 years (Rohmeder, 1941). Five good seed years were observed for cembran pine stands in the Engadin, Switzerland, between 1920–1950 giving a recurrence every 4–8 years. In these years cembran pine carried a full mast right up to timberline (Campell, 1950). According to Baig (1972) *Picea engelmannii* produces seed at timberline in the Rocky Mountains of Alberta every 4–5 years. Cones at timberline, however, usually contain undeveloped or no seeds.

Table 1. Number of spruce *(Picea abies)* seed ($\times 10^6$ ha^{-1}) in different years and altitudes from Switzerland (from Kuoch, 1965)

	1958/59 Very good seed year		1962/63 Good seed year
Spruce-Larch-Cembran Pine Forest Upper subalpine zone 1,990 m	0.4	Spruce Forest Upper subalpine zone 1,990 m	2.3
Spruce Forest Lower subalpine zone 1,600 m	12.0–25.0	Spruce Forest Lower subalpine zone 1,600 m	3.0–10.0
	1942/43 Very good seed year		
Spruce Forest Upper montane zone 975 m	22.2		

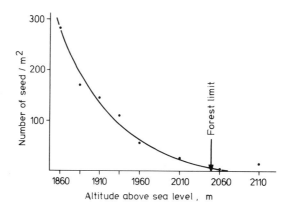

Fig. 5. Seed fall of *Picea abies* on to the snow surface in late autumn and winter at a range of altitudes. Data were collected in a subalpine spruce forest near Davos, Switzerland (from Fischer et al., 1959)

Seed years in larch, *Larix decidua*, are less frequent, due to regular destruction of flowers by late frosts and the periodic outbreak of larch leaf-roller (*Semasia diniana*) infestations impairing reproduction (Holtmeier, 1974).

Seed year interval at alpine timberline is only slightly longer than at lower altitudes, whereas the situation at the polar timberline is quite different with intervals prior to the 1920's of 60–100 years and since then, as a result of a general climatic improvement, every 10 years (Renvall, 1912; Holtmeier, 1970, 1974).

Kuoch (1965) studied seed production with respect to altitude (Table 1). He found for spruce timberline at 1990 m that in good seed years 0.4–2.3 million seed were produced per hectare. Down at 1600 m altitude, seed number per hectare increased to 3–25 million. The exponential decline with altitude in number of spruce seed dropping onto the snow surface near timberline is illustrated in Figure 5.

According to Nather (1958) cembran pine at 900 m a.s.l. developed large, symmetrical cones containing on average 100 fertile seeds. As a result of incomplete fertilisation, cones at timberline were asymmetrical in shape and contained only 10–50 seeds or even less.

Kuoch (1965) found no increase in the proportion of empty spruce seeds from 1970 m to 2090 m, but there is no doubt that the germination capacity generally declines rapidly with increasing altitude near the upper limit of tree distribution. Investigations by Holzer (1973) showed that spruce seed in the Seetal Alps, Austria, not only declines in weight from 900 mg/1000 seed at 1000 m a.s.l. to 500 mg/1000 seed at 1700 m (timberline), but also declines in germination capacity to values of $< 5\%$ at timberline. Elsewhere in the Alps, Lüdi (1938) had determined a germination rate of only 4.5% for spruce at 2000 m. At timberline in the Rocky Mountains of Alberta, Baig (1972) similarly discovered seed with very low germination percentages in *Pinus albicaulis*, *Picea engelmannii*, *Abies lasiocarpa* and *Larix lyallii*. Seed of these species, germinating at the very highest (krummholz) limit thus probably originate from the zone below timberline.

J. Wardle (1970) has studied in considerable detail the seed production of mountain beech (*Nothofagus solandri* var. *cliffortioides*) which forms timberlines extensively in New Zealand. Good seed years occur regionally on average only every 10 years, though flowers are produced more often. The amount of seed-fall and its soundness was demonstrated to decline with increasing altitude. In the good mast year of 1967 an estimate of sound seed-fall at 1000 m was over 20 million ha^{-1}, whereas at the 1350 m timberline the figure dropped to less than 0.9 million ha^{-1}, these figures representing a soundness of total seed-fall of 57% and 7% at the two altitudes respectively. In poorer mast years with some sound seed-fall at lower altitudes there is often no viable seed production at timberline. The weight of sound seed was also shown to decrease with altitude from 5.5 g/1000 seed at 1000 m altitude to 4.9 g at the 1350 m. Actual germination percentages in the field under the parent canopy were found to decrease with increase in altitude. Germination of sound seed was 20.2% at 1000 m, 9.7% at 1200 m, and only 5.1% at 1350 m after the good seed-fall in 1967.

1.2 Dispersal of Seed

The heavy wingless seed of cembran pine (*Picea cembra*) are distributed solely by animals and mainly by the nutcracker (*Nucifraga caryocatactes*), a bird which makes seed caches for winter stores. Some of these caches are not relocated, and subsequently the seed therein germinates. Since most of these winter stores are deposited above timberline, there is a continuous source of seed for natural regeneration in the timberline ecotone (kampfzone). As hiding places for seed, the nutcracker prefers certain sites such as ridges, slopes, and rock ledges. It avoids depressions such as gulleys and hollows where greater snow accumulation hinders winter access. Natural regeneration of this tree species thus advances upwards dominantly along ridges (Fig. 6). A similar situation involving another nutcracker species (*Nucifraga columbiana*) acting as an important distributor of a high-altitude five-needled pine with large seeds (*Pinus edulis*) has been described by Vander Wall and Balda (1977) for mountains in northern central Arizona.

Fig. 6. Slope with westerly aspect above forest limit in the Gurglertal, Tyrol. *Pinus cembra* is confined to the ridges and completely absent from the intervening depressions. This results from the nutcracker, *Nucifraga caryocatactes*, preferentially storing seed as a source of winter food in the more accessible ridges with less winter snow-pack

Regeneration density of cembran pine diminishes with increasing distance upwards and away from the timberline. In the Gurglertal, Tyrol, tree seedlings were found up to 250 m above the last seed-bearing trees (Oswald, 1956). Holtmeier (1966) observed occasional nutcrackers to an altitude of 400 m above timberline.

Spruce seed dispersal by wind above timberline is by contrast surprisingly poor. Trials in Switzerland produced a seed catch of 231 m^{-2} at ground level in the forest (1970–2010 m a.s.l.) and this count declined rapidly above the sharp timberline to a mere 7 m^{-2} in the timberline ecotone at 2010–2090 m. Thus the supply of spruce seed to this transition zone above timberline is very sparse in the studied region (Kuoch, 1965).

1.3 Maturation of Seed

Schmidt-Vogt (1964) has produced data of cone and seed ripening for spruce (*Picea abies*) and pine (*Pinus sylvestris*) in the Bavarian Alps. Cones of spruce mature to a water content of 40% one week later at 1100 m and 3 weeks later at 1600 m than at 600 m a.s.l. Time of seed maturation is likewise delayed at high altitude and germination percentage never attains the high values of sites at low altitude (Table 2).

Table 2. Water content of cones and germination % of seed of spruce *(Picea abies)* from different altitudes (from Schmidt-Vogt, 1964)

Time of cone collection	% water content of cones			Seed germination %		
	600 m	1,100 m	1,600 m	600 m	1,100 m	1,600 m
3.–10. Sept.	50.6	59.4	68.4	68	30	0
20.–24. Sept.	51.7	57.7	64.3	88	93	0
6.– 8. Oct.	44.7	52.5	61.9	94	94	15
20. Oct.	23.4	37.9	55.0	61	92	73
28. Oct.–3. Nov.	24.4	31.2	49.7	86	80	77
16. Nov.	18.9	20.5	33.3	96	96	84

Additionally the rate of germination of fully matured seed with a high germination percentage is slower in seed from high altitude.

The late cone and seed ripening at timberline results in spruce seed being released from cones long after there is a continuous snow cover on the ground. The seed overwinters on the snow, and then reaches the soil ready to germinate, after imbibing water during the snow-melt period. Seed in the transition zone above timberline is exposed to such extreme conditions that the actual germination percentage in early summer after the snow-melt is only 10% of the previous late autumn value (Kuoch, 1965).

Cembran pine presents a rather special case. In mature cones with seed of adequate germination capacity the seed embryo will not have attained its full length of 6–7 mm. The higher the altitude of the site, the less the embryo is developed. At a similar state of cone maturity, which occurs at 1900 m one month later than at 900 m, the mean embryo length was 4 mm at 900 m and 0.7–1.5 mm at the timberline altitude of 1900 m (Nather, 1958).

1.4 Germination

Little is known about light, moisture, and temperature requirements of germinating seed from different timberline species growing at high altitude. The limited available data would suggest these requirements are similar to those at low altitudes. Kamra and Simak (1968) found in studies with pine *(Pinus sylvestris)* of different origin that the germination percentage in seed from high altitude (1000 m) is reduced above 20° C less than seed from low altitude (300 m). It is tempting to see in this an adaptation to the high surface temperatures experienced at high altitude. Seed of *Pinus aristata* at timberline in the White Mountains (California) show optimum germination between 20–25° C (Wright, 1963). This also falls within the optimum range of lowland *P. sylvestris*. Patten (1963) investigated the effect of light and temperature on germination of *Picea engelmannii* from 2850 m altitude in Colorado and found an optimum temperature range of 15–27° C (cf. Kaufmann and Eckard, 1977) which agrees with results for *Picea abies* from the European regions of its main natural occurrence (Rohmeder, 1972, p.170).

The best pre-treatment for embryo development in cembran pine seed and its subsequent germination is when temperature is close to 20° C and the substrate holds more than 50% of the moisture at saturation (Nather, 1958). Germination percent was markedly reduced at 12° C, especially if the seed was stored cool during embryo development. On the other hand the type of substrate (sand or raw humus) played no role in germination results. Lumbe (1964) was able to increase germination percent of cembran pine by exposure to UV-radiation.

The rate of germination (germination velocity) of larch seed increases with rising temperatures between 5–30° C, but the germination percent remains constant. Not until temperatures step outside this range does the germination percent strongly decline. Larch seed can readily withstand 40° C for a short time but an exposure to 50° C for 2 h daily terminates germination. Light accelerates germination, but a dry substrate (e.g., sand with < 5% water content) impairs germination considerably (Röhrig and Lüpke, 1968).

Direct seeding trials with larch and spruce carried out in June under natural conditions at timberline in the Gurglertal, Tyrol, showed that 60–80% of the seed sown had germinated before autumn. Within the following 10 years, however, the mortality rate of the seedlings reached 90–96% (Stern, 1972).

Seeding trials with *Picea engelmannii* gave even poorer results in forest clearings near timberline (3230 m) on the Rocky Mountains of Colorado. Up to the first autumn after sowing, only 1.5% (north slope) and 0.4% (south slope) of the seed developed into seedlings with chances of survival. High seedling mortality may have been due to desiccation, animal clipping, frost-heaving, and heat-girdling (Noble and Alexander, 1977).

The ability of seed of some species to germinate at very low temperatures can be detrimental to regeneration on mountain sites with deep snow. Franklin and Krueger (1968) found seed of certain conifers, e.g., *Abies lasiocarpa* and *Tsuga mertensiana*, to germinate and perish on the surface of wet snow in the Cascades of N.W. America where heavy snow deposits are common. This may well be a contributory factor for the absence of regeneration on sites where snow lies for a long time.

1.5 Vegetative Propagation

At timberline a reproduction cycle usually stretches over several years. Individual phases of this cycle can often be checked or even completely interrupted by outside factors. Regeneration of timberline stands is then delayed for years. This situation is particularly common for spruce (*Picea abies*). However, spruce also regenerates at timberline vegetatively (literature review Kuoch and Amiet, 1970). Lateral branches of older trees develop adventitious roots, resulting eventually in circular groups of trees or timber atolls (Baig, 1972). The perimeter of such tree colonies is not always circular, but often oblong, stretching away from the prevailing wind (Fig. 7).

Kuoch and Amiet (1970) ascribe the frequent formation of spruce groups in the timberline ecotone to vegetative propagation by layering. These workers pointed out that such regeneration is not possible in a closed stand below timberline because of die-back and self-pruning of the lower branches.

Larch (*Larix decidua*) also layers, but decidedly less frequently than spruce, since the deciduous branches are not so commonly weighed down by snow to make contact

Fig. 7. Tree group or 'timber atoll' of *Picea abies* in the timberline ecotone (kampfzone) at 2100 m on Steinacherjoch (Tyrol). Such groups originate vegetatively by layering. Adventitious roots form where branches touch the litter and humus eventually allows distal parts of branches to develop into erect trees. The perimeter of these groups is usually oval in shape, extending in the direction away from the prevailing wind

with the soil (Plesník, 1973). Arno and Habeck (1972) found that *Larix lyallii* in western North America reproduces almost exclusively through seed at timberline even in extreme situations.

Islands of cembran pine, however, are not the result of vegetative reproduction. These grow from seed where the nutcracker has usually buried heaps of 10–30 seeds (Campell, 1950).

Tree groups have a decided ecological advantage over isolated individual trees in their fight for existence at timberline, as has been demonstrated by Kuoch and Amiet (1970). A tree group generates its own internal climate with ameliorated extremes, thus producing more favourable growing conditions for the individual tree.

2. Growth of Trees at Timberline

Tree form is subject to striking change with increasing altitude above sea level, especially in species occurring from valley floor to timberline. Change in life form of trees reflects growth differences primarily induced by alterations in climate. Genetically based adaptations must, however, also be taken into consideration.

In order to clarify how climate influences growth, the rate and periodicity of growth, as well as total growth increment, have been compared in uniform plant material planted out at various altitudes. If one wishes to eliminate substrate and edaphic differences, then potted plants are established at different altitudes. Genetically induced growth differences can readily be determined by testing provenances or clones under uniform conditions in growth chambers, nurseries, or homogeneous trial sites.

Growth of trees is closely dependent on their production of organic matter. Insufficient primary production will restrict growth. Conversely weak growth can also reduce dry matter accumulation, as is the case when leaf growth is impeded by climatic factors, thus allowing the photosynthetic apparatus to increase only slowly.

Under very unfavourable conditions at high altitude sites and in years with shortened growing seasons it happens that some of the annual developmental processes can no longer be completed. Plant tissues are then unable to mature fully and are damaged in the following winter.

2.1 Height Growth

Spruce (*Picea abies*) height growth in the Seetal Alps, Austria, proceeds at a mean annual rate of 33 cm in the valley at 700 m but only 8 cm in the zone above timberline at 1900 m (Holzer, 1967, 1973). This infers an average reduction in height growth of 2 cm per 100 m of altitude increase. The decline in growth above 1600 m, i.e., near timberline, is particularly rapid (Fig. 8).

Spruce seedlings established at four sites with different altitudes in the Wipptal (Tyrol) made more growth at 1250 m in all seasons than at other altitudes (Fig. 9). Less increment at the lowest altitude of 850 m was due to supra-optimal temperatures and periodic soil drought. Height growth in 1977 at 1900 m was only 20% of the maximum value found at the optimum altitude of 1250 m (Tranquillini et al., 1978).

Reduction in rate of height growth with increasing altitude in trees of comparable age is illustrated in Figure 10. Seventy-five-year old spruce in the French Central

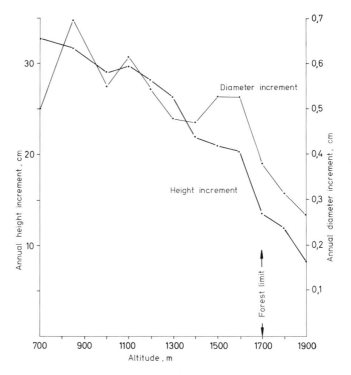

Fig. 8. Mean annual height and stem diameter increment of *Picea abies* in the Seetal Alps, Austria. 20 mature trees (70–140 years old) were selected at each altitudinal interval of approximately 100 m and mean growth rates were computed from total tree height or diameter and age. Spruce forest-limit lies at 1700 m, but isolated spruce with rapidly declining growth rates occur in the *Pinus cembra/Larix decidua* stands above this limit (from Holzer, 1973)

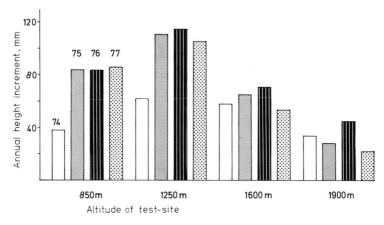

Fig. 9. Mean annual height growth of 50 spruce *(Picea abies)* clones in four consecutive seasons (1974–1977) at four altitudes in the Wipptal, Austria. Plants were established as 3-year-old rooted cuttings in 1973 (from Tranquillini et al., 1978)

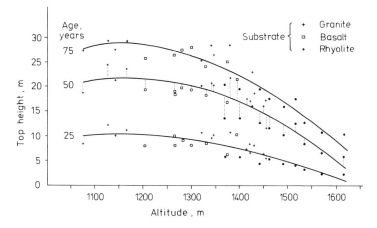

Fig. 10. Height of dominant spruce *(Picea abies)* trees in three age classes (25, 50, and 75 years) with respect to altitude and substrate in the Central Massif, France. Timberline is at 1750 m (from Oswald, 1969)

Massif reach a top height of 30 m at an altitude of 1100–1200 m and only 10 m at 1600 m (Oswald, 1969). Tree height is also dependent on the site substrate, but an altitudinal height reduction can be determined for each soil type.

Similarly the canopy height of silver beech *(Nothofagus menziesii)* in Fiordland, New Zealand, dropped from 30 m at 330 m a.s.l. to 7.2 m at 820 m (Mark and Sanderson, 1962). The top height of mature mountain beech *(Nothofagus solandri)* stands in the Craigieburn Range, New Zealand has been stated by J. Wardle (1970) as 22.1 m at 900 m, falling to 13.6 m at the 1350 m timberline.

Däniker (1923) found in Oberwallis, Switzerland, that mean spruce height declined only slightly between 1500 and 1900 m, but above timberline at 2000 m there is a very rapid height reduction to the ultimate dwarf krummholz at 2130 m. This rapid reduction is at least in part due to damage to the terminal bud and shoot (Fromme, 1963).

No change in the height of mature trees greater than 30 cm b.h.d. is detectable for larch *(Larix decidua)* or spruce *(Picea abies)* up to 1800 m on sunny aspects and 1900 m on shady aspects in the Lötschertal of the Bernese Alps, Switzerland. Above these altitudes, however, tree height declines by 3–5 m with every 100 m increase in altitude (Ott, 1978).

On Mt. Washington in Nevada, tree height also declines rapidly in the upper zone where the forest stand opens out (Lamarche and Mooney, 1972). *Pinus longaeva* (syn. *P. aristata*) still attains a height of 6 m at 3400 m but only 80 m higher at the extreme upper limit maximum tree height is 2 m.

A comparison in height growth of young cembran pine at 1985 m within the belt of high forest, and at 2190 m in the timberline ecotone (kampfzone) was made by Oswald (1963) in the Ötztal (Tyrol). The results in Table 3 show a distinct reduction in height increment within the first few years of growth over this altitudinal difference of only 205 m.

Table 3. Height of young *Pinus cembra* trees with
respect to age at timberline and in the timberline
ecotone (Ötztal, Tyrol) (from Oswald, 1963)

Age in years	Height (cm)	
	High forest (1,985 m)	Timberline ecotone (2,194 m)
5	9	4
10	21	12
15	40	25
20	63	45
25	101	73
30	147	110

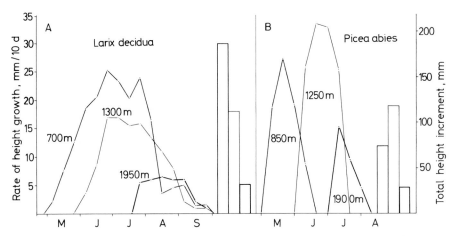

Fig. 11. Seasonal pattern of height growth and final length of terminal shoot increment
of *Larix decidua* **(A)** and *Picea abies* **(B)** at three altitudes (Tyrol, Austria). Larch of
a high-altitude provenance was potted as 1-year-old seedlings and growth was measured
the year following establishment at the various altitudes (from Tranquillini and
Unterholzner, 1968). Spruce cuttings of the early-flushing lowland clone 17 were planted
at three altitudes near Pfons, Tyrol, and assessed for growth as 4–5 year plants. Less
growth at 850 m compared to 1250 m was due to lack of moisture and weed competition
at the lowest site (from Oberarzbacher, 1977)

   The decrease in shoot increment at high altitude rests on a shortening of the
growing season and the reduction in rate of growth. This is demonstrated in trials with
potted larch seedlings (*Larix decidua*) placed at three altitudes (Tranquillini and
Unterholzner, 1968). Commencement of shoot growth is delayed the higher the
altitude, but growth terminates at approximately the same time in all altitudes
(Fig. 11A). The assumption that trees at higher altitudes are adapted to utilise the
shorter growing season particularly intensively appears false. In fact the rate of shoot
extension declines even more sharply with altitude than the length of the growth period
(Table 4A). The length of the shoot extension period was 82% at 1300 m and 52% at

Table 4A. Height increment in cm, growth period in days, and rate of growth in mm per 10 days of leading shoots of young larch *(Larix decidua)* at various altitudes. Relative values as % of data from 700 m a.s.l. (from Tranquillini and Unterholzner, 1968)

	Absolute values			Relative values		
	700 m	1,300 m	1,950 m	700 m	1,300 m	1,950 m
Ht. increment — cm	18.7	11.2	3.2	100	60	17
Growth period — days	154	127	78	100	82	51
Mean growth rate	12.2	8.8	4.1	100	72	34
Max. growth rate — mm/10 days	25	17	6.5	100	68	26

Table 4B. Height increment, growth period and rate of growth of terminal shoots of spruce *(Picea abies)* cuttings. Mean values from 14 clones of low altitude origin (from Oberarzbacher, 1977)

	Absolute values			Relative values	
	850 m	1,250 m	1,900 m	1,250 m	1,900 m
Ht. increment — cm					
1975	9.6	13.2	3.0	100	23
1976	8.5	12.5	5.2	100	42
Growth period — days					
1975	35	39	27	100	69
1976	36	32	29	100	91
Mean growth rate — mm/10 days					
1975	27.2	33.7	11.0	100	33
1976	23.4	39.1	18.1	100	46

1950 m compared to 700 m, whereas the mean growth rate per 10-day interval declined to 72% and 34% respectively, and the maximum growth rate even more to 68% and 26%. The seasonal height increment of only 60% and 17% of that at low altitude was thus demonstrably influenced even more by the reduction in rate of growth than by shortening of the period of growth.

Detailed studies of height growth of spruce *(Picea abies)* using clonal rooted cuttings were carried out at four altitudes in the Wipptal near the Brenner Pass, Austria (Lechner, 1975; Lechner et al., 1977; Oberarzbacher, 1977). In this species shoot extension also starts progressively later with increasing altitude (Fig. 11B) but unlike larch it is completed in a relatively short period. Spruce belongs to the "*Quercus*" growth type, where height growth proceeds in a single burst only to be terminated by endogenous factors after a short period (Hoffmann and Lyr, 1973). The length of the shoot extension period at timberline is thus considerably less reduced than in larch, but the rate of growth declines similarly (Table 4B).

The main difference in comparing the deciduous larch with spruce, therefore, lies in spruce completing shoot extension much earlier at low than at high altitude. The new spruce shoots and needles thus have much less time to develop and mature at timberline than in the valley.

A comparable result to that in spruce was obtained by Wardle (1972) with mountain beech (*Nothofagus solandri*) seedlings brought to a range of altitudes in the Craigieburn Range, New Zealand. Seedlings grew at all altitudes for approximately the same length of time, but start and finish of shoot extension were 6 weeks later at 1650 m compared to 1100 m. Below timberline (1350 m), seedlings had enough time to mature new growth and develop terminal buds, but above timberline they approached winter incompletely ripened and were then largely destroyed.

Other intensive altitudinal growth studies have been completed in this mountain range of New Zealand. Wardle, J. (1970) measured a seasonal mean terminal shoot increment on saplings of natural mountain beech regeneration of 40.5 cm at 820 m and 4.9 cm at timberline (1350 m). From low altitude (30 m) to timberline, start of bud growth was delayed by 2 months from the beginning of October to early December and continued till mid-autumn (late March) at all altitudes, ceasing only slightly earlier at high compared to low altitude. The growth period in these young trees was thus considerably longer at low than at high altitude.

Growth patterns of other timberline species introduced for reforestation purposes have also been studied intensively in Craigieburn Mountains (Benecke et al., 1978). Whereas all conifer species (including *Pinus contorta*, *Pinus mugo*, *Picea engelmannii*, and *Larix decidua*) showed a similar delay in start of growth of 1–2 months (depending on the season) from 900 m to timberline at 1350 m, the delay in bud-set was only 2 to 3 weeks. Growth sequence differed in the species studied. Pines and spruce rapidly completed the bulk of their shoot growth in early summer, followed by completion of needle growth. Larch, however, extended needles first and then photosynthesised for up to 2 months before extending shoots vigorously. Larch shoot growth continued until autumn as in mountain beech. Correlations with screen air temperatures at different altitudes showed shoot extension began when the mean weekly temperature began to exceed 5° C, but rapid shoot growth was not possible until the temperature rose above 7°–10° C. Comparison of growth and temperature data from a cool with those from a warm spring showed that these cardinal temperatures explain the 6-week difference between the two seasons in start of shoot growth at timberline. The lower the altitude, the less pronounced was this season-dependent delay in start of growing season.

The phenology of very young seedlings in their first two seasons of growth is not necessarily identical to that of older established trees. 2-year old Engelmann spruce seedlings, for instance, flushed 1–2 weeks earlier than young trees, and with increasing altitude, seedlings showed an increasing delay in setting a dormant bud (Benecke and Morris, 1978). The period of young seedlings without a resting bud, when shoot extension is possible, thus increased drastically with altitude. This applies to most conifers tested, so that using a Scots pine (*P. sylvestris*) provenance as an example, the seedling period without a resting bud was 40 days at 36 m altitude, 58 days at 900 m and 102 days at 1350 m (timberline). A comparison of 2-year old pine data from three different seasons at timberline confirmed that the poorer the growing season in respect to temperature, the longer the young seedlings struggled to complete their shoot extension and the setting of a dormant bud. Very young seedlings are thus exceptionally vulnerable at high altitude to shoot damage.

Benecke (1972) studied plants of *Pinus mugo*, *Alnus viridis*, *Picea abies*, and *Nothofagus solandri* var. *cliffortioides* grown at a range of altitudes up to timberline near Innsbruck, Austria. All species showed a delay in commencement of shoot activity of about 6 weeks from valley (650 m) to timberline (1950 m). The length of the shoot extension period within each species, except alder, was broadly similar at all altitudes, giving a corresponding altitudinal lag in cessation of shoot extension. The growing period for *Nothofagus*, however, was longer than the short burst of growth by spruce and pine, so that there was inadequate time for completion of shoot growth at the conifer timberline of 1950 m. Green alder appears to be less locked into an endogenous rhythm once flushed, and continues to utilise the more favourable conditions at lower altitudes, giving it a much longer period for shoot extension in the valley compared to timberline.

The short period of shoot extension (larch) and the late termination of shoot growth (spruce) in trees at or above timberline (Fig. 11) leads to the new plant tissues entering winter in a less matured state than at lower altitudes. This can be confirmed by determining the saturation water content which decreases with increasing maturation. Tranquillini (1974) found that the higher the altitude and the shorter the growing season in which spruce shoots developed, the greater was saturation water content in late autumn prior to winter desiccation. Differences in degree of shoot maturity are indicated even more clearly by the thickness of cuticula and the epidermal cutin layers of the terminal shoots and needles. According to Platter (1976), newly formed larch shoots from various altitudes showed anatomical differences in October after termination of further development. Foremost, the thickness of the cutin layers significantly decreased with increasing altitude. The same is true for needles of *Picea abies* and *Pinus cembra*, especially after a short and cool summer (Baig and Tranquillini, 1976; Table 5).

A reduction in thickness of the cutinised layers diminishes the diffusive resistance to water vapour. This significantly affects the water relations during winter of trees above timberline, and will be discussed in greater detail in Chapter 5.

Little is known to date of the influence of individual site factors on growth of trees at timberline. Turner (1971) has analysed growth with respect to local site variation of larch and spruce growing on a deforested NE slope between 2000–2200 m in the Dischmatal near Davos, Switzerland. Height increment increased with increasing radiation, especially at low wind speed. High wind velocity retards increment to the extent of eliminating the radiation effect. Under very weak radiation, however, wind up to 2 m s^{-1} is necessary for growth. These effects result from the influence of radiation and wind on the heat balance and water relations of the plants (Fig. 12).

Hellmers et al. (1970) studied temperature effects on growth of *Picea engelmannii* and found the optimum growth for seedlings in a phytotron at 19° C day temperature combined with a high night temperature of 23° C. Extreme fluctuations, e.g., 25° C/3° C or 35° C/7° C (day/night), were less favourable to growth, but plant development was quite satisfactory at constant temperature. Night temperature appeared to be the most important temperature and in a day/night cycle of 15° C/3° C,

Table 5A. Thickness of cuticular layers (μm) of current year *Picea abies* and *Pinus cembra* needles, and *Larix decidua* shoots along an altitudinal gradient from valley floor (1,000 m) to forest limit (1,950 m) and tree limit (2,100 m) on Patscherkofel near Innsbruck. Differences are significant at $P=0.001$ (from Platter, 1976)

Picea abies	Valley	4.7±0.045
	Forest limit	4.0±0.025
	Treeline	2.9±0.017
Pinus cembra	Forest limit	3.7±0.026
	Treeline	3.0±0.018
Larix decidua	Valley	5.0±0.075
	Forest limit	4.6±0.055
	Treeline	3.9±0.056

Table 5B. Thickness of cuticle and cuticular layers (μm) of *Picea abies* needles formed in 1971 (warm, long summer) and in 1972 (cool, short summer) at various altitudes on Patscherkofel, Innsbruck. Differences are significant at $P=0.01$ (from Baig and Tranquillini, 1976)

	1971	1972
Valley (1,000 m)	6.2 ±0.44	6.0 ±0.5
Forest limit (1,940 m)	6.26±0.65	5.37±0.51
Treeline (2,090 m)	4.46±0.54	4.12±0.4
Krummholz limit (2,140 m)	4.7 ±0.87	3.54±0.72

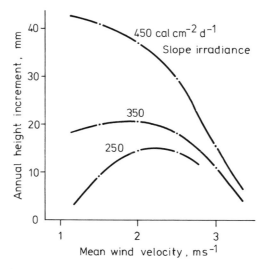

Fig. 12. Influence of mean wind velocity and solar radiation (total daily slope irradiance under a cloudless sky) during the growing season on annual height increment of young larch *(Larix decidua)* stands at 2050–2180 m a.s.l. in the Dischmatal near Davos, Switzerland (from Turner, 1971)

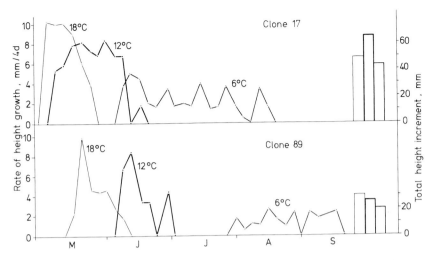

Fig. 13. Height growth of terminal shoots in mm/4 days and total height increment in mm for potted spruce *(Picea abies)* cuttings grown in growth chambers at constant temperature of 18°, 12°, and 6°C under natural light. Clone 17 of lowland origin commenced growth early while clone 89 of high-altitude origin commenced growth late in the season. The plants were brought into the chambers on 15 April (from Oberarzbacher, 1977)

which corresponds approximately to mean timberline temperatures during the growing season, height growth was only 1% of the growth at optimum temperatures (19°C/23°C). Slow growth of trees at timberline may thus be primarily dependent on low night temperatures. This study shows that though Engelmann spruce grows naturally in cool climates (Morgenthal, 1955), it may not be well adapted to the temperatures of its natural range (mean July temperature 13°C).

Temperature influence on height growth of clonal spruce *(Picea abies)* plants in growth cabinets was investigated by Oberarzbacher (1977). Low temperatures considerably delayed the breaking of bud dormancy (Fig. 13) especially in clones with late bud-break, which require more heat as a precursor to growth. Rate of shoot growth also declined with temperature, but total increment at the lowest temperature was not seriously reduced because growth continued for a longer period under the otherwise favourable conditions of the growth cabinets. In the field the time available for development of shoot primordia in the terminal bud plays an important role in predetermining the height growth of spruce the following season. Needle primordia in the terminal bud developed between mid-July and mid-September in 1976 at low altitude (Fig. 14). The number of primordia grew rapidly to a total of 115 per bud. At timberline (1950 m) the terminal bud did not set until mid-August, so that needle primordia could not begin to develop before the end of August, and further development was halted in mid-October with the coming of winter. In addition, cooler weather at timberline compared to lower altitudes ensured that only 44 primordia were formed per bud. The number of needle primordia reflected the following season's shoot increment of 130 mm at low altitude and only 13 mm at high altitude (Unterholzner, 1978).

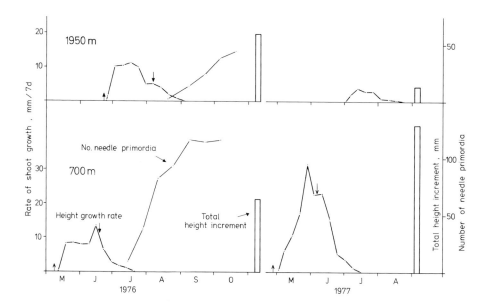

Fig. 14. Weekly increment and seasonal sum of height growth as well as number of needle primordia in the terminal bud for 3-year-old spruce *(Picea abies)*, origin Kitzbühl, Austria (1200 m) placed in spring 1976 at a valley site (700 m) and at timberline (1950 m). Growth differences were small until the second season (1977) after transplanting (from Unterholzner, 1978)

Time of bud-burst in spring is also photoperiodically determined. Whereas warm treatment at 15° C in a glasshouse will advance bud activity of cembran pine shoots by 3 months from June to March, long-day treatment (16 h) advances it by 7 months to November (Schwarz, 1970). Time of bud-burst is, however, also dependent on an endogenous annual rhythm. The readiness of timberline larch shoots to flush when transferred to favourable temperatures diminishes in winter from December to February and increases again thereafter (Machl, 1969). In December it took 13 days for needles to develop to a length of 5 mm at 20° C and natural day length. In February this period was 20 days, in March 15 days, and in April only 6 days. The slow reactivation in spring of the ability for timberline larch to flush is a consequence of low temperatures which counteract the rising activation readiness induced by the endogenous rhythm and increasing day length.

Growth differences in trees at different altitudes are also genetically based (Rohmeder, 1964). High altitude origins generally flush earlier when grown on the same sites (Münch, 1923; Nägeli, 1931) and have slower growth rates than low altitude origins, but a later time of bud-break of high elevation provenances has also been found (Vincent and Vincent, 1964; Štastny, 1971; McGee, 1974). Rohmeder (1964) established seedlings of a lowland (800 m) and a high altitude (1800 m) larch *(Larix decidua)* provenance together at three different altitudes, namely 700, 800, and 1150 m. Seedling height growth was followed for 10 years (Table 6). Both origins declined in height growth with increasing altitude of test site, and though the lowland

Table 6. Mean height in cm at age 10 years of a high and low altitude provenance of *Larix decidua* at three different sites (from Rohmeder, 1964)

	High altitude origin (1,800 m)			Low altitude origin (800 m)		
	700 m	800 m	1,150 m	700 m	800 m	1,150 m
Altitude of test site (m a.s.l.)						
Mean height of trees (cm)	179	114	96	271	148	127

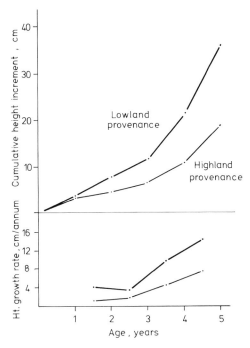

Fig. 15. Mean total height and annual height increment of spruce *(Picea abies)* seedlings of two origins from the Seetal Alps, Austria during the first 5 years of growth. Seed of lowland and highland provenance trees were collected at 800–1500 m and 1500–1600 m respectively (from Holzer, 1966)

origin grew fastest at all sites, its superiority in this respect declined with altitude. This suggests that the high-altitude provenance is better adapted to growth under high altitude conditions than the lowland origin whose greater growth potential cannot be fully realised at high altitude.

Holzer (1966) raised spruce seedlings under uniform conditions at 500 m from seed collected from trees at various altitudes. Again height growth of seedlings of high-altitude origin was decidedly less than that of low-altitude seedlings (Fig. 15), partly because the former commenced shoot extension later and formed at terminal bud sooner. Earlier growth termination of high-altitude seedlings is due to a response at a longer photoperiod, rather like long-day plants. Further progeny were propagated vegetatively from these spruce and planted at various altitudes in the Wipptal, Tyrol.

Table 7. Height increment, shoot extension period and rate of growth of terminal shoots in spruce *(Picea abies)* cuttings from different origins and established at three altitudes in the Wipptal above Pfons, Austria. Mean of 14 lowland and 20 high altitude provenances together with differences between these in % of the lowland value are presented (from Oberarzbacher, 1977)

	850 m	1,250 m	1,900 m
Mean increment (mm)			
Highland provs. 1975	78.7	99.5	26.6
Lowland provs. 1975	96.0	131.8	29.8
Difference in %	− 18.0	− 24.5	− 10.7
Highland provs. 1976	81.0	106.5	40.8
Lowland provs. 1976	84.6	124.8	51.5
Difference in %	− 4.3	− 14.7	− 20.8
Mean period of extension (days)			
Highland provs. 1975	35.8	33.9	24.6
Lowland provs. 1975	35.3	39.1	26.9
Difference in %	+ 1.4	− 13.3	− 8.6
Highland provs. 1976	30.8	27.0	26.6
Lowland provs. 1976	36.1	31.9	28.5
Difference in %	− 14.7	− 15.4	− 6.7
Mean growth rate (mm day^{-1})			
Highland provs. 1975	2.20	2.93	1.08
Lowland provs. 1975	2.72	3.37	1.11
Difference in %	− 19.1	− 13.1	− 2.7
Highland provs. 1976	2.63	3.94	1.53
Lowland provs. 1976	2.34	3.91	1.81
Difference in %	+ 12.4	+ 0.8	− 15.5

A comparison of mean growth for two seasons showed clones of high-altitude origin to put on less growth at all planting sites than clones of lowland origin (Table 7). This growth inferiority is primarily due to a shorter shoot extension period and less due to a slower rate of extension, which can in fact at times be greater in the high-altitude origin (Table 7). Thus there was no sign in the first three years after planting of the high-altitude origins performing better in respect to height growth at high-altitude than lowland origins.

The crown form in spruce is almost certainly genetically fixed (Holzer, 1967). Whereas at low altitude one finds dominantly broad club-shaped crowns, the high-altitude ecotypes are characterised by slender cylindrical crowns (pointed or columnar) (Fig.16). This is probably an adaptation to snow conditions, since slender crowns carry less snow, thereby reducing the danger of branch- and leader-breakage (Mlinšek, 1973).

It is also possible for crown form to have evolved in response to radiation climate. Slender crowns predominate at polar timberlines with little snow-fall (Holtmeier, 1974), perhaps because within a stand such forms enhance utilisation of light incident from low solar altitude (Schmidt-Vogt, 1977).

Fig. 16. Spruce *(Picea abies)* stand at 1800 m a.s.l. near the Jaufenpass, Italy. Trees show narrow columnar crowns typical for high-altitude sites and are perhaps an adaptation to snow-fall

2.2 Growth of Leaves

Many studies have shown length and area of needles and leaves to decline with increase in altitude above sea level. Cembran pine needles attained a length of 7.6 cm at 1300 m and only 5.2 cm at the 2000 m timberline (Tranquillini, 1965b). In larch, needle growth does not diminish with altitude to the same degree as shoot growth, nevertheless the reduction in needle length from 700 m to timberline at 1950 m was from 3.5 cm to 1.9 cm. For needles of larch the altitudinal delay in start of shoot growth is partly compensated by the extended growth period and high maximum growth rates.

Needle shortening and reduction in leaf area with climbing altitude was also reported by Benecke (1972). This effect was particularly marked in green alder *(Alnus viridis)* with individual leaf area of 20 cm^2 at 650 m in the valley and 6 cm^2 at the 1950 m timberline.

Besides the shortening of needles with altitude for which further examples could be cited (Holzer, 1967; Müller-Stoll, 1954; Platter, 1976), the other dimensions of width and thickness not unexpectedly also decrease, as was determined for spruce (Holzer, 1973). Needles, however, are more crowded about the twig axes at high altitude sites. The number of stomata per unit length of needle showed no significant difference between altitude of site (cf. Želawski and Niwiński, 1966). The increased life of needles with altitude is noticeable. Whereas spruce needles can attain an age at 300 m a.s.l. of 4–5 years on the tree bole and 5–7 years on lateral shoots, at high altitude sites these ages reach 9–10 years and 11–12 years respectively (Holzer, 1967). The ash content of needles was also found to diminish with increasing altitude.

Such changes in needle structure can be shown to exist in well-developed trees up to timberline. The trend then ceases in the transition zone above timberline, as demonstrated by Müller-Stoll (1954) on the Feldberg in the Black Forest, Germany. In well-formed spruce trees needle length and width decreased, but needles were thicker and more densely spaced at higher altitude. Once trees show signs of stunting further up the slope, all needle dimensions, including thickness and spacing density, rapidly decline (Table 8).

Needle thickness was also found to increase with rising altitude in *Pinus ponderosa* and *P. jeffreyi* (Haller, 1962). This may not be solely an environmental modification, but can be a genetic difference. Needles of high-altitude provenances become thinner when grown at low altitude, but they are still thicker than the needles of low-altitude provenances grown alongside.

Little is known to date of anatomical differences in needles from various altitudes. Bonnier (1895, 1920) found in plants, including conifers, grown at

Table 8. Dimensions of current year needles of spruce *(Picea abies)* on Feldberg (Black Forest, Germany) at different altitudes (from Müller-Stoll, 1954)

	Forest (1,240 m)	Tree groups of an open stand (1,350 m)	Upper krummholz zone (1,435 m)	
			3 m tree	1 m deformed bush
Needle length (mm)	18.72	12.62	12.37	8.04
Needle width (mm)	1.36	1.17	1.18	1.02
Needle thickness (mm)	0.72	0.92	0.97	0.56
Spatial density (needle No. per cm shoot length)	22.2	26.2	32.5	24.2
Current year shoot length (cm)	4.98	3.06	2.61	1.41

lowland and mountain sites that cell walls in the needle epidermis, cuticle and hypodermis were much thicker in the mountain plants. At high altitude structural features characteristic of "sun" plants began to appear (cf. Napp-Zinn, 1966).

Against this Baig and Tranquillini (1976) measured thinner cuticles as well as thinner epi- and hypodermal layers in spruce and cembran pine needles at timberline compared with needles from the valley. Platter (1976), however, found the cellulose layer in epidermal cells of spruce needles at timberline to be no different in thickness than at low altitude. In cembran pine needles at the tree limit the strongly sclerotic epidermal cells are in fact taller than epidermal cells at timberline.

2.3 Diameter Growth

Stem diameter growth decreases with altitude just as height and leaf growth (Ott, 1978). The annual breast–height radial increment of spruce in the Seetal Alps, Austria, is 6 mm at low and moderate altitudes falling rapidly above 1600 m to 3 mm at timberline (Fig. 8). Diameter increment does, however, appear to decline less with altitude than height growth (Däniker, 1923; Oswald, 1969). Various site factors influence diameter growth so that its correlation with altitude is less strong than with height growth (Fig. 8). Wardle, J. (1970) compared diameter growth in stands at 900 m with that at the 1300 m timberline and found little change in breast–height diameter growth with change in altitude within the same *Nothofagus solandri* association, though as expected large differences between associations (and sites) were recognised.

A reduction in radial growth at high altitude is related to the shortened growing period largely as a result of the delay in start of seasonal growth. Kern (1960) noted that cambial activity in spruce began 4 weeks later at 1350 m in the Black Forest, Germany, than at 230 m in the Rhine Valley. Termination of annual ring formation was independent of altitude and occurred at the same time.

In southern Norway, Mork (1960) measured a decline in diameter increment for spruce from 5.0 mm at 140 m to 1.5 mm at 860 m altitude. Radial growth terminated about the same time in mid-July, but lignification of newly formed conducting elements commenced very much later at high altitude sites. Though initially the lignified proportion of the current annual ring at 860 m is small, lignification does reach completion at the same time as at 140 m altitude (Fig. 17).

Žumer (1969) also concluded that radial growth period is shortened solely by delayed commencement of growth. Lignification of the annual ring occurs simultaneously with cell wall thickening and continues for about 2 weeks after completion of cell division. In Norway, conductive xylem elements lignify even at the highest altitude sites. It is possible, however, on extreme sites in seasons with premature arrival of winter that lignification is not completed (Sigmond, 1936).

According to Krempl (1978) commencement of early-wood cell differentiation is deferred by only 2 weeks from 900 m to 1800 m (timberline) in the Murau region, Austria, whereas for bud-burst the delay is 6 weeks. The

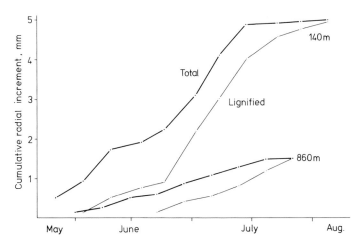

Fig. 17. Stem radial increment and lignification of the annual ring in spruce *(Picea abies)* at 140 m and 860 m in Southern Norway. Samples taken at breast height (1.3 m) from trees with 25 cm diam. b.h. (from Mork, 1960)

first late-wood cells, however, are laid down about the same time at all altitudes and termination of late-wood cell lignification occurs on average 2–3 weeks earlier at timberline than in the valley.

The time-sequence of annual ring formation in larch seedlings has been studied at three altitudes by Tranquillini and Unterholzner (1968). Cambial activity began with swelling of cells at mid-altitude (1300 m) 13 days and at high altitude (1950 m — timberline) 71 days later than in the valley (700 m). There was a corresponding delay in further stages of cambial development, e.g., beginning of cell division and differentiation of early-wood tracheids. The start of late-wood formation, though, no longer showed such large delays with altitude and cambial activity ceased simultaneously at all sites (Fig. 18).

Radial increment in the larch seedlings dropped from 1.43 mm at 700 m to 0.97 mm at 1300 m and 0.34 mm at 1950 m, but not solely as a result of the shortening growing season at higher altitudes. Rate of growth also declined so that mean rate of increment per 10-day intervals for the season was 77% at mid-altitude and 37% at timberline of that at 700 m (Fig. 18).

In 1976 spruce seedlings previously established at 700 m began cambial activity between 10th and 20th April. (Oberarzbacher, 1977). The first early-wood cells were formed at the beginning of May and formation of late-wood cells began towards the end of June. Cambial cell division ceased between mid and late August, thus giving a radial growth period of 130 days (Fig. 19). At 1950 m activity began 35–50 days later, but continued for 11–18 days longer at the end of the season, so that the mean radial growth period was only 20 days shorter than in the lowland.

There are references to the cambium entering dormancy earlier at high than at low altitudes (e.g., Daubenmire, 1945). According to Ermich (1960) growth-ring formation at 1000 m begins 2 weeks later in fir *(Abies alba)* and 4 weeks later in beech *(Fagus sylvatica)* than at 520 m in the Tatra Mountains,

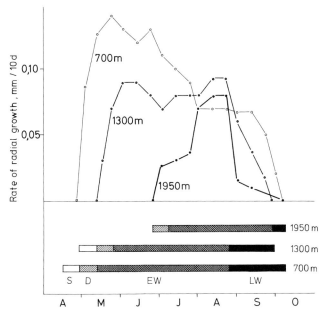

Fig. 18. *Upper*: Seasonal sequence of radial growth (ring width increment in mm per 10 days) for 2-year-old larch *(Larix decidua)* at three altitudes in Tyrol, Austria. *Lower*: Phases of cambial growth: Swelling of cambial cells *(S)*, cambial cell-division *(D)*, early wood *(EW)* and late wood *(LW)* formation of plants at 700, 1300, and 1950 m (from Tranquillini and Unterholzner, 1968)

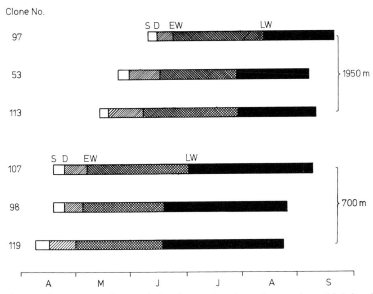

Fig. 19. Seasonal sequence of annual ring formation, i.e., phases of cambial development, in 4–5-year-old potted spruce *(Picea abies)* cuttings of different clones at low (700 m) and high (1950 m) altitude near Innsbruck, Austria. *S*: Swelling; *D*: Cell division; *EW*: Early wood formation; *LW*: Late wood formation. Clones originate from sites between 800–1500 m. The differences between clones in commencement of cambial activity are similar to differences in start of shoot growth (from Oberarzbacher, 1977)

Czechoslovakia. In spite of this delay, cambial dormancy was attained 2 weeks earlier at the higher altitude. The percentage late-wood of the total growth ring thus declines wite increase in altitude.

Decrease in diameter growth increment with altitude must be dominantly caused by the corresponding decline in temperature. At cooler temperatures there is a tendency for photosynthate to be transformed to sugars and starch rather than cellulose, and this can limit diameter growth. This is the reason why fluctuations in annual ring width occurring in the mountains and at high latitudes sensitively reflect small changes in mean growing season temperatures (Hustich, 1948; Lamarche and Fritts, 1971; Bednarz, 1976). If summers favourable to growth are rare, then trees grow very slowly (Billings and Mooney, 1968).

Not only does the annual timber volume increment decrease with rising altitude (e.g., Wardle, J., 1970) but timber quality also deteriorates. Vorreiter (1937),when studying the causes of snow breakage, discovered that spruce wood loses in specific density as well as in bending and compression strengths as one approaches closer to the timberline on the Glatzer Schneeberg, Silesia. Damage through snow breakage in the mountains thus not only depends on increased snowfall at higher altitude sites, but also on weaker structural strength of the wood.

The lignin content of spruce wood fell from 29.8% at 410 m to 26.5% at 1450 m and to 23.8% in krummholz (Zinn, 1930). Cieslar (1897) had already presented similar results very much earlier. This reduction in lignin content not only affects timber strength, but also its durability. Timber from higher altitudes is more readily decomposed by fungal enzymes (Gäumann, 1948; Gäumann and Péter-Contesse, 1951).

Exposed trees in the ecotone above timberline often build compression wood in response to mechanical pressure from wind or snow and to one-sided crown development (flagged trees; Däniker, 1923). This has a detrimental effect on the conductive function of the wood.

2.4 Root Growth and Mycorrhizae

Seasonal root growth, like other growth processes, begins much later on the mountains than down in the valleys, but it terminates at all altitudes more or less at the same time (Tranquillini and Unterholzner, 1968). Root development was interrupted several times during the growing season in potted larch seedlings growing at low altitudes; only at the highest site of 1950 m (timberline) was uninterrupted growth of root tips recorded (Fig. 20). This is comprehensible in terms of Hoffman's finding (1972) that root growth depressions and cessation during the growing season are primarily a consequence of soil drought which decreases in frequency with increasing altitude.

Root growth initiates as soon as soils become free of frost and rise above a certain soil temperature minimum (Bannan, 1962; Bode, 1959). This minimum threshold temperature for root growth of *Pinus taeda* lies at a diurnal thermocycle between $10°C$ and $2°C$ and $10°C$ and $4°C$ (Bilan, 1967). After a period of

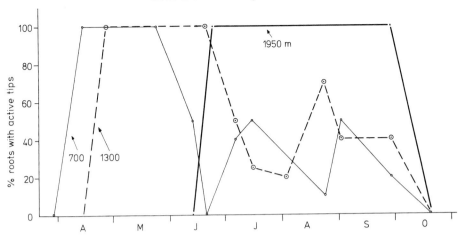

Fig. 20. Root activity during the growing season expressed by the proportion (%) of the total root tips which were actively growing in 2-year-old potted larch *(Larix decidua)* placed at three altitudes in Tyrol, Austria (from Tranquillini and Unterholzner, 1968)

dormancy between early December and beginning of March, root growth was reactivated in the field in eastern Texas when the minimum air temperature remained above 0°C and the maximum exceeded 21°C.

In cool soils, however, root growth is very slow. When seedlings of larch and spruce were transplanted at a soil temperature of 4°C, virtually no new roots were produced within the first month. At 12°C the root regeneration potential increased slightly but did not approach optimum until soil temperature was 20°C (Tranquillini, 1973). In view of the fact that soil temperatures down to 1m depth at timberline near Obergurgl, Tyrol, lie constantly below 5°C for 6 months, and even in the growing season 15°C is never attained at 20cm depth (Aulitzky, 1961a), then we may draw a general conclusion that tree roots can only grow very slowly at high altitudes in the European Alps.

Root studies described by Benecke et al. (1978) in the milder mountain climate of New Zealand showed a different situation. Even at timberline (1300 m) soil temperature in 10cm depth did not fall below 0°C, and during the summer of root measurement mean soil temperatures were above 10°C for 4 months. Rapid rates of root extension were measured in the field at 900 and 1300 m with a timberline mean maximum in *Pinus contorta* lateral roots of over 3 mm/day. The period for active root extension was at least 8 months for *P. contorta* and *Nothofagus solandri*, and root dormancy for these species at timberline was estimated at less than 2–3 months. It was concluded that root systems must be capable of physiological activity for 9 months or more of the year.

The root–shoot ratio of plant dry weight gave an indication of the relative decline with altitude of root growth in potted spruce and mountain pine *(Pinus mugo)* established at several altitudes (Benecke, 1972). Data in Table 9 show that at high altitude root growth decreased even more strongly than shoot growth compared to lower-altitude sites. A similar altitudinal decline in root : shoot ratio was also demonstrated for one-year seedlings of *Nothofagus solandri*

Table 9. Ratio of root: shoot dry weight in 4-year-old
plants at three altitudes (from Benecke, 1972)

	650 m	1,300 m	1,950 m
Picea abies	0.76	1.03	0.33
Pinus mugo	0.82	0.67	0.38

(Wardle, 1971) and 2-year-old seedlings of *Picea engelmannii* (Benecke and Morris, 1978) over a range of altitudes in the Craigieburn Range, New Zealand.

An essential prerequisite for tree root systems to be optimally effective at timberline appears to be intensive development of mycorrhizae. Wardle (1971) pointed to the necessity of this if seedlings of such timberline species as *Nothofagus solandri*, *Eucalyptus pauciflora*, *Picea engelmannii*, *Pinus contorta*, *Pinus flexilis*, and *Pinus hartwegii* are to establish sucessfully near timberline.

Investigations of mycorrhizae in cembran pine occurring at central European alpine timberlines have shown that even at time of vigorous root growth a maximum of only 2.3% of short roots are nonmycorrhizal and possess root hairs (Göbl, 1967). The assumption by Baig (1972) that above a certain altitude mycorrhizal development is no longer possible due to low soil temperatures and therefore induces a timberline would certainly appear not to apply to the alpine timberline formed by cembran pine. That mycorrhizal fungi have also evolved high-altitude strains adapted to low temperatures was demonstrated by Moser (1958).

Moser (1967) maintains that only trees with ectotrophic mycorrhizae are capable of reaching the upper forest zones, and that timberlines would be up to several hundred metres lower if the mycorrhizal symbiosis did not exist. In his opinion it is only the combination of tree and ectotrophic fungus which can gather adequate nutrients for growth and maturation of new shoots in sites with short growing seasons such as at timberline.

3. Dry Matter Production of Trees at Timberline

In the last chapter it was shown that growth of trees declines as one approaches the timberline, resulting in reduced stature and size of plant organs. This fall-off in growth could underlie an increasing shortage of organic materials, itself a consequence of deficient primary production. It is thus important to look closely at altitudinal changes in the primary production of trees. The magnitude of primary production in part determines the ability to compete with other plants and species, as well as the resistance to limiting environmental factors. Primary production, therefore, is of central importance in the occurrence of plants on specific sites.

If we look at the schematic outline of the processes involved in primary productivity of a tree at timberline, then the focal point is the total seasonal net photosynthesis of the tree's foliage (Fig. 21). Net photosynthesis is reduced

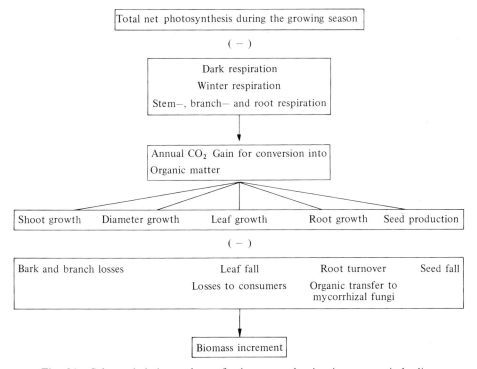

Fig. 21. Schematic balance sheet of primary production in trees at timberline

by night respiration of leaves, winter respiration of the whole plant, and respiration of all photosynthetically inactive tissue. The balance presents the tree's annual CO_2-gain available for conversion into organic matter. This is then utilised for the tree's material increment and is laid down in the various plant organs. The tree, however, also loses organic matter through die-back and shedding of plant parts, by exudation of organic substances, and to consumers of plant material. The balance between production and all losses gives the biomass increase of the tree.

It appears relevant to investigate the magnitude of dry-matter production of trees under the extreme conditions of the timberline ecotone (kampfzone), and to determine if growth increment is limited by a deficiency in available organic matter. According to Boysen-Jensen (1949) the carbon balance at treeline can even be negative, so that trees are then unable to develop and they eventually die.

3.1 Photosynthesis

Photosynthesis is of prime importance for dry-matter production in plants. Its magnitude is largely dependent on both the situation of the moment and the previous environmental history, whereby most external factors exercise a specific influence on photosynthesis. With the advent of infrared gas analysis in 1950 it has been possible in recent decades to clarify and quantify the important relationships between photosynthesis and climatic factors by (a) simultaneous monitoring of CO_2 gas-exchange and environmental factors in the field and (b) measurement of photosynthesis in growth chambers by varying one factor at a time while keeping others constant (Tranquillini, 1968). These investigations have allowed one to identify in detail how trees at timberline react to the various climatic factors, what significance these factors have for life of trees under limiting conditions, and the extent to which photosynthesis of timberline trees is adapted to environmental extremes.

3.1.1 Dependence of Photosynthesis on External Factors and Their Significance for CO_2-uptake of Trees at Timberline

3.1.1.1 Light

At alpine timberline light conditions for photosynthesis are less favourable than might be expected from the increase with altitude of incoming radiation. This is foremost due to timberlines generally following contours on steep slopes where elevated horizons screen the site from the total possible insolation. The mean daily radiation total at timberline on a slope with westerly aspect at 2000 m in the Gurglertal, Austria, is only 10–20% higher in summer than at 200 m in the lowland (Turner, 1961a).

In order to determine the response of net photosynthesis under natural field conditions to changing radiation, CO_2-uptake of cembran pine *(Pinus*

Table 10. Regions of irradiance and light (absolute values and in percent of measured maximum values), in which net photosynthetic rate of young cembran pine *(Pinus cembra)* at the timberline was negative, light-limited, optimal or supra-optimal under natural site conditions (from Turner and Tranquillini, 1961)

	Radiation intensity cal cm^{-2} h^{-1}	Illuminance K Lux	Net photosynthesis rate	% of Maximum
I	0 – 1.2	0 – 1.6	Negative	0 – 1.1
II	1.2–18	1.6–25	Light-limiting	1.1– 16.8
III	18 –42	25 –60	Optimum	16.8– 39.9
IV	42 –107	>60	Limited by supra-optimal temperature and hydro-active stomatal closure	39.9–100

Table 11. Duration of different levels of radiation intensity at timberline during the growing season. Mean for 1954 and 1955 in h and % of total possible sunshine duration (from Turner, 1961a)

Total	<1.2	1.2–18	18–42	>42	cal cm^{-2} h^{-1}
2,351	162	955	621	613	h
100	6.9	40.6	26.4	26.1	%

cembra) regeneration at timberline in the Gurglertal was tracked simultaneously with the irradiance intensity. The summarised results illustrate a typical optimum curve because high radiation in the field is coupled with high leaf temperatures and low relative atmospheric moisture levels. We can distinguish several response phases (Table 10).

Isolated young cembran pines cannot utilise radiation intensities greater than 42 cal cm^{-2}h^{-1} (60,000 lux illuminance) for increasing rates of net-carbon gain, in fact such intensities restrict net production.

Turner (1961a) was able to calculate the time duration of light intensities in each of the four response phases (Table 11) during the course of the growing season (16 May–31 October). Light was optimal for only 26% of daylight hours, supra-optimal for a similar length of time and for the remainder, i.e., almost half of the time available, light was limiting at timberline.

The daily sum of net photosynthesis in young cembran pine did not decline until the daily radiation total dropped below 150 cal cm^{-2}d^{-1}, i.e., 19% of maximum possible radiation. Even this rather low threshold radiation total was not attained on 14% of all days during the May–September period on an open site at timberline.

The light regime just below timberline within cembran pine/larch stands with canopies approaching closure was considerably poorer. The mean radiation near the ground of a densely regenerating site was only 9.6% of the radiation measured in the open (Turner, 1958a). Cembran pine regeneration and shade branches of tree crowns are well adapted to the low light levels existing under stand canopies. Photosynthesis balances respiration (i.e., light compensation

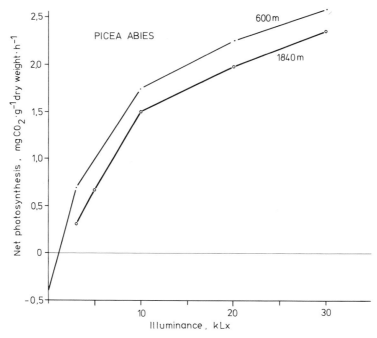

Fig. 22. Net photosynthesis at $15°C$ of 1st- and 2nd-year excised sun shoots from mature spruce *(Picea abies)* trees in the Botanical Gardens, Innsbruck (600 m) and at timberline on Patscherkofel (1840 m) as a function of illuminance from a Xenon lamp (from Pisek and Winkler, 1959)

point) at $0.1 \, cal \, cm^{-2} h^{-1}$ and maximum net photosynthesis is already reached at $12 \, cal \, cm^{-2} h^{-1}$ (Tranquillini, 1955). For shade-tolerant foliage, total daily photosynthesis is thus not limited, unless radiation totals fall below $48 \, cal \, cm^{-2} d^{-1}$. In spite of this shade-tolerance, plants and branches within the stand rarely achieve optimum rates of CO_2 assimilation. On dull days during the growing season radiation sums are just sufficient for cembran pine seedlings to compensate total plant respiration (Turner and Tranquillini, 1961).

These findings would indicate that the opening-up of high altitude stands commonly seen near timberline is to the advantage of primary production because of increased crown illumination. The question whether trees from other altitudes have a similar light dependence is best answered by investigation of a species with a broad altitudinal range, e.g., spruce *(Picea abies)*. Pisek and Winkler (1959) produced light-response curves for spruce twigs taken from adult trees at 600 m and 1840 m altitude. Up to the maximum of 30,000 lux used in this investigation the foliage from near timberline always gave a lower assimilation rate, based on needle dry weight, than foliage from the valley. The shape of the response curve showed no clear difference, however, and there is little to suggest that timberline trees are able to make better use of high light intensities (Fig. 22; cf. Fryer et al., 1972; McNaughton et al., 1974).

Other spruce species, e.g., *Picea engelmannii* in the Rocky Mountains, apparently also show no special adaptation to high light intensities and light saturation is attained at 40,000 lux (Ronco, 1970). Even cembran pine, the timberline species so characteristic for the central European Alps, reaches light saturation at 25,000 lux (Tranquillini, 1955). The low light optimum and the photosynthetic ability to exploit weak light label the species mentioned so far as shade plants, at least at the seedling stage. Their inability to adapt to the light climate at high altitude was demonstrated for *Picea engelmannii* at 2700 m, where photosynthesis of planted seedlings lagged well behind seedlings growing under artificial shade (Ronco, 1970). Perhaps this sensitivity to strong light is the cause for the tendency of forests to reach their upper limit as closed stands, at least where shade species are involved. Wardle (1974) has used a similar argument to help explain the abrupt timberline of *Nothofagus solandri*, which demands shade as a seedling at high altitude in New Zealand.

In contrast to these results, photosynthesis of *Pinus contorta* seedlings at 2700 m increased to 120,000 lux, and no detrimental effects on photosynthesis were noted from exposure to full sunlight (Ronco, 1970). Such sun-species tend to form very open timberline stands (Walter, 1968).

3.1.1.2 Temperature

a) Short-Term Response

When measuring photosynthesis continuously at timberline in the Gurglertal, needle temperatures of cembran pine seedlings were also recorded. This gave the first insight into the temperature dependence of CO_2-uptake at a field site. The response curve showed a surprisingly low optimum temperature for photosynthesis between 10–15° C (Tranquillini, 1955).

Others subsequently carried out laboratory investigations into the temperature dependence of photosynthesis in various types of plants including timberline species (Pisek and Winkler, 1959; Pisek and Rehner, 1958; Pisek et al., 1967, 1968, 1969).

The minimum temperature for CO_2-uptake by cembran pine in the field in late autumn was −4.7° C (Tranquillini, 1957). Since needle water begins to freeze at −4° C (Tranquillini and Holzer, 1958) the lower temperature compensation point for photosynthesis is most probably dependent on ice formation and the associated dehydration of the protoplasm (Pisek et al., 1967)

Leaves of plants at timberline show a slight but distinctly lower temperature mimimum for photosynthesis than leaves of similar age and development in the valley. In *Vaccinium* species this difference was 1.5° C, and in larch needles *(Larix decidua)* taken from trees at 900 m and 1900 m in late summer it was 0.5° C (Pisek et al., 1967). Even this minor shift in the temperature curve is of considerable advantage to vegetation at high-altitude sites where weak summer frosts are frequent (Tranquillini and Holzer, 1958).

Differences in photosynthetic temperature optima between mountain and valley plants of the same species are more pronounced. In spruce *(Picea abies)* from timberline, the temperature optimum was 3° C lower than spruce from the valley floor, irrespective of light intensity. Larch *(Larix decidua)* and birch *(Betula verrucosa)* (Fig. 23) gave similar results (Pisek et al., 1969). Such data

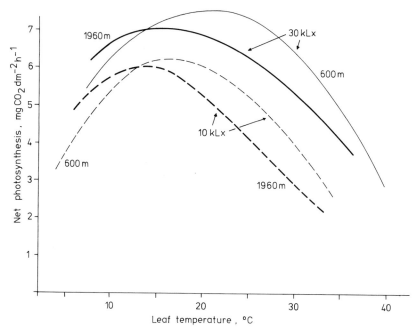

Fig. 23. Net photosynthesis (for total leaf area, i.e., upper and lower surface) of excised twigs from mature birch *(Betula verrucosa)* trees in the Botanical Gardens, Innsbruck (600 m) and the krummholz limit in the Sellraintal, Tyrol (1960 m) as a function of temperature at two levels of light (10 and 30 klux Xenon light) (from Pisek et al., 1969)

clearly demonstrate an adaptation to the cool climate at timberline. The earlier decline in net photosynthesis with rising temperature and the generally lower photosynthetic temperature maximum (36° C in *Pinus cembra*) of timberline trees can be explained at least in part by the fact that trees respire more strongly at timberline than at lower altitude (cf. Chap. 3.2). Thus at equal gross-photosynthetic rates temperature optimum and maximum would be expected to shift towards lower temperatures.

Adaptation of photosynthesis to cool climates is largely a labile process. If plant individuals of the same species and origin are maintained for several weeks at both low and high temperatures, then plants are produced with correspondingly lower and higher temperature optima (Mooney and West, 1964; Mooney and Shropshire, 1967; Neilson et al., 1972; Smith and Hadley, 1974; Slatyer, 1977). *Pinus aristata* plants brought down from timberline, where the temperature optimum under an illuminance of 90 klux lies between 10–15° C, to Los Angeles increased their temperature optimum for photosynthesis to 20° C. Respiration was simultaneously and drastically reduced (Mooney et al., 1966).

For the Australian timberline species, *Eucalyptus pauciflora*, Slatyer and Morrow (1977) demonstrated a parallel shift in photosynthetic temperature optimum with the seasonal change in mean air temperatures. This results at any point in time in the temperature optimum of plants at the cooler high-altitude site being lower than that for plants growing at warmer low altitudes. Additionally

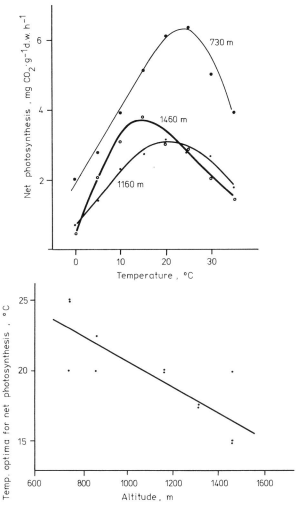

Fig. 24. *Upper:* The dependence of net photosynthesis on temperature at 22 klux illuminance for seedlings of *Abies balsamea* raised in a glasshouse from seed collected at various elevations in the White Mountains, New Hampshire (USA). *Lower:* Regression of optimum temperature for net photosynthesis on elevation of origin for Balsam Fir seedlings (from Fryer and Ledig, 1972)

at low temperature, the photosynthetic rate of trees at the highest altitude was greater than the rate at lower altitudes. The reverse was true at high temperatures and confirms that there must also be a genetic component in this adaptation.

There are other examples where photosynthetic acclimation to low temperatures is at least partly genetically determined. Fryer and Ledig (1972) measured photosynthesis at a range of temperatures in seedlings of *Abies balsamea* raised from seed collected at various altitudes in the White Mountains, New Hampshire. The higher the altitude of seed origin, the lower was the temperature optimum. The lapse rate in optimum was 2.7° C per 300 m altitude (Fig. 24) and corresponds to the lapse rate in mean temperature maxima for the summer months.

Similar results were obtained by Slatyer (1977) with *Eucalyptus pauciflora* plants raised from seed collected at 915 and 1770 m and grown in a phytotron. Plants from high-elevation seed showed a temperature optimum of 20° C when grown at a 15/10° C regime and 25° C when grown at a 33/28° C regime. By comparison, photosynthetic optimum for the low elevation seed source was 25° C when grown at 15/10° C and this shifted to 30° C when grown at 33/28° C.

A number of plant species representative of high-altitude sites in the White Mountains, California, were shown to have a lower photosynthetic temperature optimum than lowland species when grown under identical conditions (Mooney et al., 1964; West and Mooney, 1972). Against these findings Oberarzbacher (1977) was unable to demonstrate any significant difference in temperature optimum between high- and low-altitude clones of spruce (*Picea abies* from 1500–1600 m and 500–900 m) whether grown under uniform conditions or at the divergent mountain and valley climates.

In comparing the frequency distribution of actual needle temperatures with the photosynthetic temperature-response curve of *Pinus cembra* (Tranquillini and Turner, 1961), one sees that the most common needle temperatures always

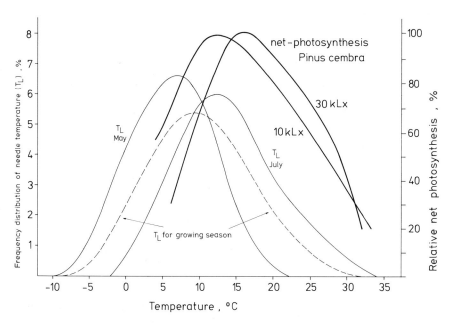

Fig. 25. Dependence of net photosynthesis on temperature at two levels of light for *Pinus cembra* shoots (from Pisek and Winkler, 1959). Frequency as percent of total daylight time of needle temperature (T_L) on a *Pinus cembra* seedling at timberline for the whole growing season as well as for May and July. Except at low light (10 klux) during the warmest month (July), needle temperatures for the greatest proportion of time are too cool to coincide with optimum CO_2-uptake (from Tranquillini and Turner, 1961)

lie below the temperature optimum for 30 klux. Only in the warmest seasonal month of July did greatest temperature frequency coincide with the photosynthetic optimum temperature for low light levels, i.e., 10 klux (Fig. 25). Needles of cembran pine at timberline in the middle of the growing season, but particularly early (May) and late (September) in the season, are seldom warm enough for optimum CO_2-assimilation.

b) Persistence of Frost Effects and Induction of Winter Dormancy

It became obvious in field measurement of photosynthesis at timberline during autumn in the Gurglertal, Austria, that the daily total CO_2-uptake by cembran pine *(Pinus cembra)* was always well below the expected mean on days following a night with frost (Tranquillini, 1957). The check imposed by frost on the photosynthetic apparatus appears to show subsequent persistence. The extent of this persistence is directly related to the severity and frequency of the frosts experienced by the trees (cf. Pisek and Kemnitzer, 1968; Bauer et al., 1969). This leads to a step-wise decline in total daily CO_2-uptake during early winter in the field (cf. Chap. 3.1.2.1). Photosynthetic capacity finally ceases completely and not even under optimum temperature and light is respiration fully compensated (Pisek and Winkler, 1958).

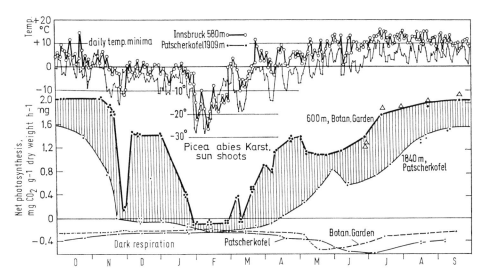

Fig. 26. Seasonal differences between net photosynthetic capacity, and dark respiration, (mg $CO_2 g^{-1}$ d.w. h^{-1}, 12°C, 10 klux) of sun shoots of *Picea abies* in the Botanical Garden, Innsbruck (600 m) and at the timberline, Patscherkofel (1840 m). Seasonal course of daily temperature minima is given for weather stations at both sites. In the relatively mild valley climate photosynthetic capacity was negative only during the coldest month (February), whereas at the colder timberline net photosynthesis remained below compensation point for 5 months (December–April). The early summer depression in curves for both sites was due to new developing shoots initially respiring very strongly (from Pisek and Winkler, 1958)

The fall in photosynthetic capacity induced by frosts below $-4°C$ sets in at timberline long before doing so at lower elevations, since such frosts come earlier and with greater frequency at high altitude (Fig. 26).

Dormancy in cembran pine is, however, not solely induced by cold temperatures. Photosynthetic capacity of this pine fell from October through to January, then increased thereafter, even under a constant temperature of $15°C$ in a glasshouse. This activity cycle is predominantly controlled by photosynthetic capacity is presumed to be controlled by an endogenous rhythm. when day-length is also maintained constant. The remaining weak change in photosynthetic capacity is presumed to be controlled by an endogenous rhythm. This is supported by cembran pine covered for long periods at $0°C$ and near darkness by deep snow. Such plants still showed some seasonal change in photosynthetic capacity, frost hardiness, and degree of dormancy (Schwarz, 1968).

c) Termination of Dormancy

At the point of deepest winter dormancy, photosynthetic capacity is not only shut down, but respiratory activity is also reduced. Desiccation and frost resistance are at a maximum, whereas the potential for new growth and reactivation of photosynthesis at warm temperatures is at a minimum. The gradual termination of this dormancy proceeds more slowly and later at timberline than at lowland sites. The photosynthetic capacity of spruce *(Picea abies)* became positive at the beginning of March in the valley but at timberline compensation point was not exceeded until mid-April (Fig. 26). Similar investigations in the Rocky Mountains of Alberta with several tree species at 1400 m elevation showed photosynthetic reactivation to occur between the end of March and end of April, but at timberline (2300 m) *Abies lasiocarpa* was not reactivated until mid-May (Schwarz, 1971).

The factors responsible for increasing photosynthetic activity after breaking of dormancy are the same as those controlling the transition into dormancy, predominant among them being temperature. If cembran pine seedlings are kept at a constant $15°C$ in the glasshouse, then photosynthesis begins as early as January and attains summer rates in March (Bamberg et al., 1967). In the field, however, tree foliage exposed above the snow-pack at timberline remains inactive considerably longer. Frosts are here still frequent well into May and repress the reactivation process.

A strict photosynthetic dormancy appears to be a characteristic of tree species growing in low-temperature climates, particularly at timberline. For a more thermophilic species such as *Picea sitchensis*, Neilson et al. (1972) found greatest frost sensitivity in autumn (September/October) when the photosynthetic capacity in the field could be reduced to zero by night frosts. This sensitivity was temporary and rapid frost hardening (2–4 weeks) subsequently occurred, increasing steadily to mid-winter. In early winter (December) the photosynthetic capacity of hardened shoots continued unreduced after frost. Any photosynthetic dormancy in the relatively mild frost climate of Scotland was thus very short and transient for sitka spruce.

3.1.1.3 Wind

Besides temperature, wind is a decisive climatic factor for timberline. Its inhibitory influence on height growth of young larch *(Larix decidua)* and spruce *(Picea abies)* at sites above the forest stands has already been noted (cf. Chap. 2.1). We now consider the influence of wind velocity on photosynthesis and whether the increased wind at high altitude (Aulitzky, 1961b; Nägeli, 1971) restricts dry matter production.

Experiments in a climatised wind tunnel demonstrated that various woody species react very differently to wind (Tranquillini, 1969). Cembran pine *(Pinus cembra)* and larch *(Larix decidua)*, which occur naturally on wind-exposed sites in the ecotone above timberline (kampfzone), maintained maximum assimilation rates at moderate wind speeds and did not begin marked reduction

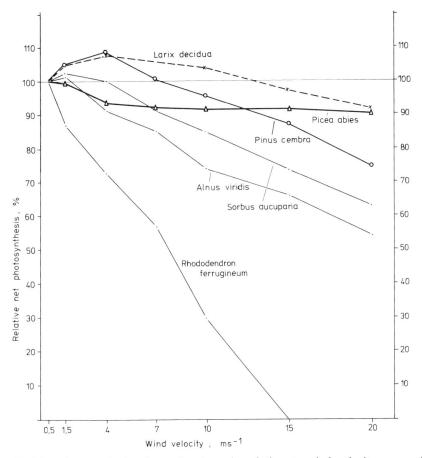

Fig. 27. Net photosynthesis of woody plants in relation to wind velocity, expressing P_N as a percentage of the value at $0.5 \, \text{m} \, \text{s}^{-1}$. Prior to experiments the potted plants were watered to saturation and then placed in a climatised wind tunnel at 30 klux, $20°\text{C}$, 12 mb vapour pressure deficit, $300 \mu l \, l^{-1} \, CO_2$ and $15°\text{C}$ soil temperature. Wind velocity was increased in steps with plants remaining at each level for 3 h (Tranquillini, 1969)

in CO_2 gas exchange until speeds of $10 \, m \, s^{-1}$ were reached. *Picea abies, Alnus viridis*, and *Sorbus aucuparia* reacted much more sensitively to rapidly moving air and their CO_2-uptake already diminished at $4 \, m \, s^{-1}$. The alpine dwarf shrub, *Rhododendron ferrugineum* tolerated even less wind and occurs naturally almost exclusively in sheltered leeward subalpine sites. Stomatal closure was already significant at $1 \, m \, s^{-1}$ wind and at $15 \, m \, s^{-1}$ closure was complete, so that net photosynthesis dropped to zero (Fig. 27). It is thus not surprising to find this species completely absent from sites exposed to wind.

Caldwell (1970a) found *Pinus cembra* maintained open stomata even in strong wind, and only the most recently formed needles reacted more sensitively by closing stomata. The observed decline in photosynthesis of this species was ascribed to needles being pressed closer together by the wind and thus receiving less light for photosynthesis (Caldwell, 1970b). This secondary effect reduced photosynthesis by only 20–40% even at high wind velocity ($15 \, m \, s^{-1}$).

Wind characteristics in the ecotone above timberline in the Gurglertal (2200 m) were investigated 30 cm above ground at three sites with different exposure to wind from June to October (Caldwell, 1970c). Wind velocity was rarely above $3 \, m \, s^{-1}$ in the wind-shelter of runnels, but on the crest of adjoining ridglets it was frequently greater than $5 \, m \, s^{-1}$, but even there velocities of $8 \, m \, s^{-1}$ or more were exceptionally rare. Photosynthesis in low bushes of *Pinus cembra* and *Larix decidua* growing close to the surface is little reduced by wind. One must, however, consider that in the field wind dries the soil surface (Neuwinger, 1961) and reduces the leaf temperature. Thus wind can impair photosynthesis indirectly.

By contrast, CO_2-uptake in free-standing tall trees near timberline is probably frequently limited by exposure of foliage to strong winds.

3.1.1.4 CO_2-Content of the Atmosphere

The volumetric concentration of CO_2 in the atmosphere decreases with altitude so that at 2000 m it is 79% and at 3000 m 70% of the value at sea level. The corresponding reduction in CO_2 partial pressure leads to a lower diffusive gradient from atmosphere to leaf and thus reduced CO_2-uptake. Theoretical considerations by Gale (1972a) indicate that diffusion of CO_2 into leaves at high elevations probably declines less than expected from the decline in CO_2 partial pressure.

Mooney et al. (1964) compared photosynthesis of various plants from the White Mountains, California, at 300 and $200 \, \mu l \, l^{-1}$ CO_2 in the laboratory. These concentrations correspond to those in free air at 930 m and 4900 m. Photosynthesis in $200 \, \mu l \, l^{-1}$ was on average 40% (30–49%) less than in $300 \, \mu l \, l^{-1}$ CO_2. This reduction in photosynthesis was similar for plants irrespective of their altitudinal origin (1340–3860 m). From this it is concluded that plants at high altitude show no adaptation to the low CO_2 concentrations at such sites (Mooney et al., 1966).

In the region of timberline, atmospheric CO_2 content is not as low as expected. In the European Alps at 1900–2600 m, the mass concentration is on average $0.38 \, mg \, CO_2 \, l^{-1}$ air as against $0.45 \, mg$ at 800 m (Pisek, 1960). These values, however, lie in the range where photosynthesis of woody species, including

Abies alba, is directly proportional to the atmospheric CO_2 content (Koch, 1969). One must thus reckon with a fall of 10–20% in photosynthetic performance of trees at timberline due to lower CO_2 levels.

3.1.1.5 Soil Temperature

The low soil temperatures at timberline (Aulitzky, 1961c) have been shown to limit dry matter production of trees and particularly to delay reactivation of full photosynthetic potential in spring (Havranek, 1972). In laboratory trials a reduction in soil temperature from 15° to 2°C induced the CO_2-uptake of *Larix decidua* and *Picea abies* seedlings to fall by 20–30%. A similar reaction was obtained for *Alnus viridis* and *Pinus mugo* (Benecke, 1972).

Dormant *Pinus cembra* seedlings brought into a warm glasshouse in March progressed to active photosynthesis slowly when roots were maintained cold (Havranek, unpubl.). After 400 h the rate of net photosynthesis was 16% less at a soil temperature of 2°C compared to a soil temperature of 8°C.

The depressive influence on production of low soil temperature was confirmed in field measurement of net photosynthesis at timberline. After emergence from under a snow cover in May, seedlings of *Pinus cembra* had very low daily photosynthetic totals when soil temperature was still close to 0°C. Daily totals then increased slowly proportional to the rise in mean daily soil temperature until 7°C was reached, above which soil temperature no longer showed a limiting effect on photosynthesis (Fig. 28).

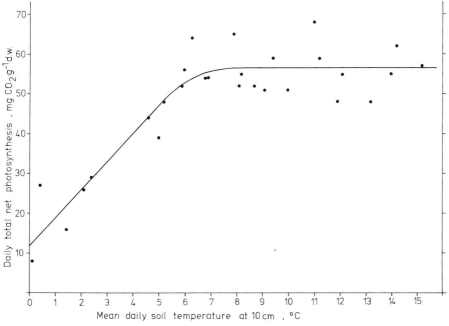

Fig. 28. Daily sum of net photosynthesis of natural regeneration of *Pinus cembra* seedlings at timberline near Obergurgl, Austria, during the periods 13 May–4 June and 2–13 July (from Tranquillini, 1959) in relation to daily mean soil temperature at 10 cm depth on the same site. Daily net photosynthetic sum increased with soil temperature between 0–7°C, above which there was no further marked increase (from Havranek, 1972)

At soil temperatures below $-1°C$, the water readily available to plants freezes and water uptake from the soil is curtailed (Larcher, 1957). When the tree root-zone freezes, stomata respond to moisture deficits with closure and CO_2-uptake ceases within a few days (Tranquillini, 1957). The seasonal time span available for net carbon-gain can thus be considerably shortened by soil frost.

Studies in the Sierra Nevada, California, produced no evidence for a significant correlation of photosynthetic sensitivity to low soil temperatures and altitude of site (Anderson and McNaughton, 1973). However, populations of *Typha latifolia* at 1900–2600 m showed better water and nutrient uptake at low temperatures than populations from sea level (McNaughton et al., 1974).

Walter and Medina (1969) have pointed out the great significance of soil temperature for the altitudinal zonation of vegetation on tropical mountains. Favourable soil temperatures on coarse screes are said to be responsible for allowing isolated stands of *Polylepis sericea* to grow in the high-alpine zone of the Venezuelan Andes to 4200 m a.s.l. Spomer and Salisbury (1968) suggested that timberlines lie at an altitude above which soil temperatures are too low for adequate growth and functioning of roots. According to the climatic and ecological data from timberline studies in the European Alps, soil temperatures are high enough, at least from June to September, not to limit growth and dry matter production critically.

3.1.1.6 Atmospheric and Soil Moisture

Many plants, including timberline conifers such as *Larix decidua*, *Picea abies*, and *Pinus cembra*, restrict photosynthesis in dry air when soil moisture is nonlimiting (Tranquillini, 1963a). Photosynthesis falls off below 50% relative humidity and at 25% gave rates of only 8% *(Picea abies)*, 40% *(Larix decidua)*, and 43% *(Pinus cembra)* of the photosynthesis measured at 80% relative humidity. Spruce is adapted to cool but moist climates and is considerably more sensitive to low humidity than the other two species (cf. Neuwirth et al., 1966). Net photosynthesis of *Picea sitchensis* shoots grown in a glasshouse were unaffected by leaf–air vapour pressure differences up to 11 mbar but photosynthesis declined with a further increase, largely as a result of increasing stomatal closure (Ludlow and Jarvis, 1971).

Since this effect is essentially a response to atmospheric water-vapour pressure deficit it declines rapidly with increasing altitude. The common mid-day depression of photosynthesis induced by moisture stress on fine days is less pronounced and occurs less frequently at high altitude. In this way the negative influence of low temperature, strong wind, and low CO_2-concentration on dry matter production of trees at timberline is ameliorated a little by the lower atmospheric moisture deficits (i.e., higher humidity).

Soil moisture also influences net photosynthesis. Seedlings of timberline species *Larix decidua*, *Picea abies*, and *Pinus cembra* started a gradual decline in photosynthesis as soon as soil water potential decreased to between -0.4 bar (pine) and -3.5 bar (larch) (Havranek and Benecke, 1978; Fig. 29). While soil potential remained above -3 to -5 bar, the photosynthetic decrease was

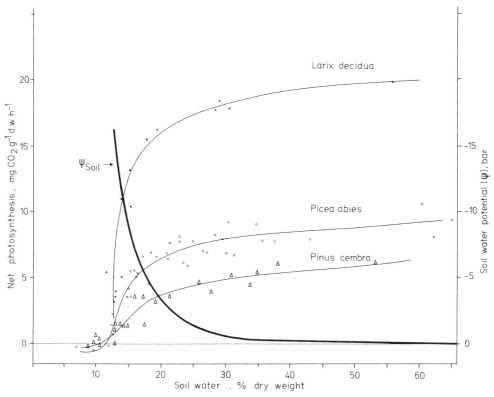

Fig. 29. Net photosynthesis of timberline tree species in relation to soil moisture content. Potted seedlings were watered to saturation and then allowed to desiccate in a controlled-climate chamber. Sample plants were periodically removed for determination of net photosynthesis at 30 klux, 20° C, 12 mb vapour pressure deficit, 2.5 m s^{-1} wind and 300 μl l^{-1} CO_2. Soil moisture, after gravimetric determination, was then used to estimate soil water potential (ψsoil) from the ψsoil/water content curve previously established with a ceramic plate extractor. Net photosynthesis began to decline at soil moisture of 35% (ψsoil −0.5 bar) and reached compensation point in all three species at 10–12% soil moisture (ψsoil −15 to −19 bar); (from Havranek and Benecke, 1978)

small, but below this potential the reduction was rapid. At a soil water potential of approximately −16 bar CO_2-uptake had completely ceased.

Near timberline, however, soils only rarely become dry enough for primary production of trees to be limited by soil moisture. Soil water potential determinations at various sites in the transition zone above timberline by Gunsch (1972) support this view. During summer at 1800 m in the Central Alps (Austria) on a deforested slope with southern aspect and low annual precipitation (1075 mm) water potential in the upper soil layers (2–5 cm) remained close to field capacity. Only during one late autumn fine weather period at the beginning of October did soil water potential drop below −2.5 bar to an absolute minimum of −8 bar. In another valley (Gurglertal) where less precipitation falls (940 mm) soil water potentials on a wind-exposed westerly slope towards the end of a

9-day rainless period fell to -7.4 bar, but on a sheltered southerly site the potential was still at -2.5 bar. At other times the soil remained moist, even in this region with exceptionally low precipitation for an alpine zone (Turner, 1970).

Thus any lengthy and decisive limitation to dry matter production in trees caused by soil drought is most unlikely to occur at the alpine timberline.

3.1.2 Annual Course of Net Photosynthesis of Trees at Timberline

Net photosynthesis, i.e., CO_2-uptake, was measured on natural regeneration of *Pinus cembra* and *Larix decidua* almost continuously for a complete year at 2070 m (timberline) in the Gurglertal, Austria. This series of measurements is unique for such an elevation and allows a first insight into the photosynthetic behaviour of trees growing in the field under the complex interaction of all environmental factors (Tranquillini 1957, 1959, 1962, 1964a).

3.1.2.1 Seedlings of *Pinus cembra* Covered by Winter Snow

As soon as cembran pine regeneration reappeared from under the melting snowpack (e.g., mid-May 1955), CO_2 gas-exchange was positive (Fig. 30). Daily CO_2-uptake totals remained fairly low at around 20 mg g^{-1} d.w. d^{-1} for the first 10 days after release from under snow. During this period photosynthesis was still limited by low soil temperatures (cf. Chap. 3.1.1.5). With rising soil and air temperatures the daily photosynthetic totals climbed threefold to 60 mg g^{-1} d.w. d^{-1} by the end of May and maintained this level till early July. A noticeable reduction in net CO_2-uptake then set in as buds burst and new growth of the trees began. This led to a broad depression in the seasonal photosynthetic curve until late August, though radiation and temperature conditions were optimum during this period and plants never lacked

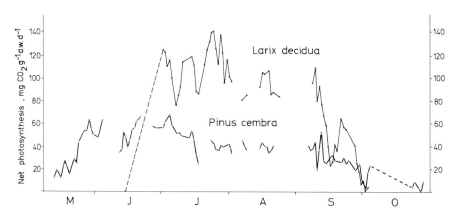

Fig. 30. Seasonal sequence of the daily sum of net photosynthesis for natural regeneration of *Pinus cembra* and *Larix decidua* at timberline near Obergurgl (2000 m), Austria. Larch commenced photosynthetic activity one month later than cembran pine and also ceased activity earlier than pine. Larch's shorter photosynthetic period is compensated for by higher rates of net photosynthesis than in pine (from Tranquillini, 1962)

soil moisture. After September there was a step-wise decline in daily net photosynthetic totals as a result of decreasing radiation and temperature, and the progressive persistence of photosynthetic inhibition caused by nightly frosts. In mid-November, after several very cold days, frost reached the root zone of the young trees, thereby interrupting the uptake of water by the plant from the soil. This quickly led to stomatal closure and cessation of CO_2-uptake.

Early in December plants became snowed-in and under the snow-cover it was too dark for any abundant carbon gain. Even at strong insolation (70 klux) *Pinus cembra* seedlings under 15 cm of snow could only assimilate CO_2 at half the expected maximum rate, and under 50 cm of snow they barely reached CO_2 compensation point (Tranquillini, 1957). Since the depth of snow was greater than 50 cm for the period 10 December to 3 May it was concluded with a fair degree of certainty that snowed-in plants fixed no significant quantities of CO_2.

In years when winter snow arrives late, the period during which soils are frozen also limits annual CO_2-uptake and dry matter production of cembran pine seedlings. At the ecological field station Obergurgl, Tyrol, snow-pack disappears on average in mid-May and starts building up again mid-November (Fig. 31). Plants thus experience a mean period of 186 days, i.e., 6 months, without snow. Release from snow varies less from year to year than the time of becoming snowed in. Early snow falls have reduced the seasonal snow-free period to 155, and late snows have extended it up to 219 days.

Above the closed forest stands in the timberline ecotone zone, wind primarily determines the depth and length of snow cover according to the surface topography of the slope. The snow-free period can thus vary greatly over short distances. In a sheltered gully, for example, *Rhododendron ferrugineum* was not released from snow until 6 June, giving a snow-free period of only 146 days. On an adjoining wind-exposed rise with an *Alectoria-Loiseleuria*-heath

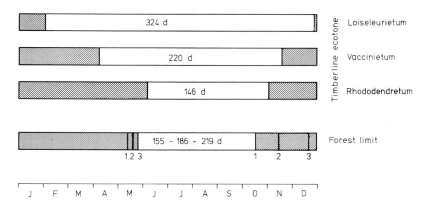

Fig. 31. Mean seasonal duration of snow cover (*dotted area*) and snow-free period (No. of days) near Obergurgl, Austria, at timberline (2070 m) and in various plant communities in the timberline ecotone (kampfzone) (2200 m) during the years 1953/54 till 1958/59. *1*, earliest; *2*, average; *3*, latest time of snow melt and arrival of new snow-cover (from Turner, 1961b)

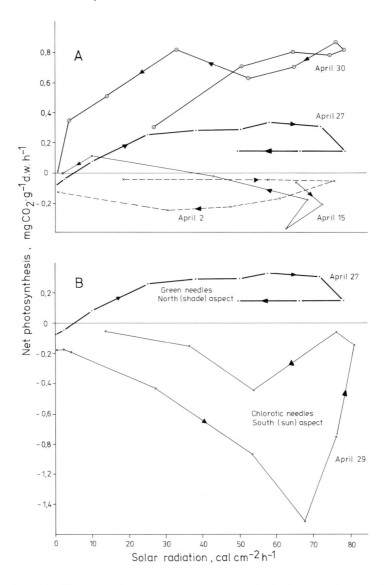

Fig. 32 A and B. Typical daily courses of net photosynthesis during spring (April, May) of *Pinus cembra* branches exposed above the snow surface in winter at timberline near Obergurgl, Austria, with respect to radiation. Hourly rates are joined in time sequence. **A** Rise in photosynthetic rates during April. Net photosynthesis, P_N, was still negative during daylight hours on the 2 April. After a considerable rise in temperature the compensation point was surpassed in the afternoon on the 15 April. P_N was weakly positive all day on the 27 April and rates of CO_2-uptake continued to rise on the 30 April. **B** Comparison of net photosynthesis in green foliage from the shady north side of the crown and chlorotic foliage from the sunny south side during late April. Net photosynthesis of the chlorotic needles remained negative during daylight. During the morning, respiration increased rapidly with rising temperatures, but in the afternoon activation of photosynthesis caused increasing reassimilation of respired CO_2 (from Tranquillini, 1958)

community snow only lies for a few weeks. In the *Vaccinium uliginosum* community where *Pinus cembra* regenerates most readily, the snow-free period was 220 days.

Secular and site differences in arrival of snow do not, however, greatly influence the annual plant production. Photosynthesis is already progressively declining in September and if snow arrives late in the season, frost halts photosynthesis prior to a snow cover. By contrast, in the spring any extension of snow coverage beyond April significantly diminishes the annual production of cembran pine seedlings, since by then they have reached a point of high potential activity and begin CO_2-uptake immediately after release from snow.

3.1.2.2 Trees of *Pinus cembra* Exposed Above the Winter Snow-Pack

Although during the winter larger trees of cembran pine stand free above the snow surface, their period of photosynthetic activity is no longer than seedlings buried under the snow. In both cases CO_2-uptake progressively declines from September onwards due to deficient light and heat, as well as due to the persistence of shut-down after frosts. In November when seedlings are snowed in or their CO_2-uptake is terminated by soil frost, then taller trees have also entered into true dormancy. Brief periods of optimum weather no longer result in CO_2-uptake (Pisek and Winkler, 1958). Winter dormancy in trees at timberline is particularly rigid and long-lasting. Between February and April there is no photosynthesis to compensate the respiration, which climbs rapidly at temperatures above $0°C$, as demonstrated by CO_2 release under light. Thus trees at timberline cannot take advantage of short warm weather periods, which regularly occur under "Föhn" conditions. There is no carbon gain and considerable losses occur instead through active respiration.

Rigid winter dormancy begins to break during April and trees can then recommence CO_2-uptake towards the end of April (Fig. 32). Needles on branches exposed in winter to strong radiation were chlorotic and became photosynthetively active later than green shade foliage.

3.1.2.3 *Larix decidua*

The duration of the photosynthetic period in deciduous trees is dependent on the length of time they bear foliage. Larch seedlings at timberline in the Gurglertal, Tyrol, unfolded new needles in 1955 not before mid-June at a time when maximum photosynthetic rates were already being recorded in *Pinus cembra*. Larch needles senesced and turned yellow in the second half of September.

The seasonal photosynthetic period for larch of only 107 days was thus considerably shorter than the period for evergreen cembran pine of 181 days in the same season. Older larch trees in the same locality at 2100 m were shown after 6 years of observations to flush on average on 6 June and fully senesce on 12 October (Fig. 33). This corresponds to a photosynthetic period of 128 days, i.e., 4 months. In regions with a lower land mass and on the shaded side of valleys the period is even shorter (Friedel, 1967).

The shorter photosynthetic season in larch compared to cembran pine is more than compensated by the higher photosynthetic capacity of larch. Whereas

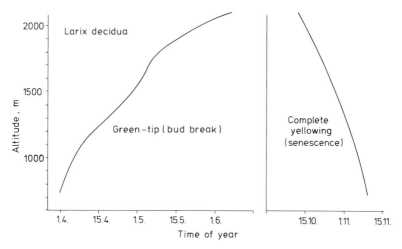

Fig. 33. Change with altitude in time of bud-break (new needles 2–3 mm) and senescence of larch *(Larix decidua)* needles in the Ötztal, Austria, between 700 and 2100 m a.s.l. Data are the mean of six seasons. Delay with increasing altitude in commencement of spring activity was much greater than the delay in autumn senescence with descending altitude. The period of photosynthetic activity for larch was thus 222 days at 700 m and only 128 days at the 2100 m timberline (from Friedel, 1967)

the maximum daily CO_2-uptake of needles was 69 mg g^{-1} d.w. for pine, the figure was 143 mg g^{-1} d.w. for larch, i.e., approximately twice the assimilation rate of cembran pine. This high rate of CO_2-uptake was attained two weeks after the "green-tip" phase and continued until the end of July (Fig. 30). In September the rapid decline in larch photosynthesis paralleled that of cembran pine, but with the break-down of chlorophyll, CO_2-uptake ceased sooner in larch.

Larch accumulated in a complete season 9.2 g CO_2 g^{-1} d.w. This is 46% more than the seasonal net gain by cembran pine (6.2 g CO_2 g^{-1} d.w.) at the same site in the same year. Larch's superior performance results from better illumination of needles in a more open crown, and from the fact that the proportion of unproductive foliage tissue is by weight much less in larch than in pine (Tranquillini, 1962).

This high rate of net photosynthesis based on foliage dry weight does not mean that mature larch trees fix more carbon than cembran pine trees of similar size. Total needle weight of a 70-year pine (height 7 m) at timberline was 8.44 kg, more than twice the needle weight of 3.37 kg for a larch tree of similar height (Tranquillini and Schütz, 1970). Total annual tree net assimilation for cembran pine with its lower photosynthetic capacity was 44.0 kg CO_2, clearly more than the comparative figure of 29.3 kg CO_2 for larch.

3.1.2.4 Other Published Investigations

The first studies of photosynthesis at alpine timberline of trees and dwarf shrubs were published by Cartellieri (1935). He worked in the Tyrolean Central

Alps on Patscherkofel near Innsbruck, and used the titration technique developed by Boysen Jensen, adapted for field use by Bosian, to determine CO_2-uptake of *Pinus cembra* on several days during the course of a year at 1900 m. At the end of March, in spite of high levels of radiation and relatively warm temperatures, no evidence of CO_2-uptake could be found. Photosynthesis became slowly active towards the end of April (daily total $9.2 \, mg \, CO_2 \, g^{-1}$ needle d.w.) and reached a level of $20.4 \, mg \, CO_2 \, g^{-1} d.w. d^{-1}$ in June. Cartellieri found CO_2-uptake to be still at summer maximum in early September ($22.4 \, mg \, CO_2 \, g^{-1} d.w. d^{-1}$). After night frosts in October, CO_2-uptake was strongly checked, and towards the end of the month it was terminated completely. Only on warm days, e.g., 29 October, was some CO_2-uptake still to be measured ($6.8 \, mg \, CO_2 \, g^{-1} d.w. d^{-1}$).

This seasonal pattern agrees well with results obtained in greater detail using more advanced techniques 20 years later. That Cartellieri's photosynthetic values were considerably lower can be ascribed to his use of shoots from adult trees which assimilate at lower rates than seedlings (Tranquillini, 1959; Tranquillini and Schütz, 1970).

Mooney and co-workers investigated in summer 1963 and winter 1965/66 net photosynthesis of *Pinus aristata* at the upper timberline of 3100 m a.s.l. in the White Mountains of California. During the summer photosynthesis in optimum light and temperature was maintained at largely a steady level between $1.0-1.3 \, mg \, CO_2 \, g^{-1} d.w. h^{-1}$ even when in July and August a period of drought prevailed with soil water potential in the root zone down at -15 bar (Mooney et al., 1966). In the very cold winter of 1965/66 mean daily air temperatures remained steadily below $0°C$ from mid-November to the end of March and extreme minima fell to $-30°C$. Net photosynthesis had markedly declined by November and from January till April it remained negative. Schulze et al. (1967) calculated the carbon balance for the winter period (1 November–23 April) at $-140 \, mg \, CO_2 \, g^{-1}$ needle dry matter. To recover this massive winter loss *Pinus aristata* with its low photosynthetic rates requires at least half the summer period available for CO_2-uptake.

3.1.3 Photosynthesis with Respect to Altitude

In order to understand the reasons for the change in dry matter production with changing altitude, it is first of all necessary to investigate how the length of the seasonal production period and the photosynthetic rates during this period relate to changes in elevation.

The photosynthetic production period is readily determined at various altitudes for deciduous trees such as larch, which occurs from valley floor to timberline. According to Friedel (1967) the period during which larch bears green needles is on average (n = 6 years) 222 days at the bottom of the Ötztal, Tyrol, (700 m) and diminishes to 128 days at timberline (2100 m) near the head of the same valley. This shortening of the photosynthetic period is largely due to the delay in commencement of shoot activity, and to a much lesser extent to earlier needle fall at high altitude (Fig. 33).

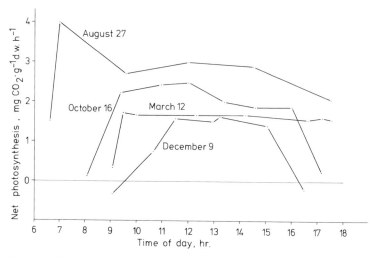

Fig. 34. Typical daily sequences in fine weather of net photosynthesis of sun twigs from a mature spruce *(Picea abies)* in the valley near Innsbruck between August and March (from Pisek and Tranquillini, 1954)

It is more difficult to determine the productive period in evergreen trees. To this end one needs seasonal patterns of photosynthesis from sites at different elevations. For *Pinus cembra* at timberline the period has been shown to be approximately 6 months. Positive CO_2-assimilation has never been shown to occur in the depth of winter at timberline, but in a milder climate at lower altitude winter photosynthesis does take place. Seedlings of cembran pine growing near Zürich, Switzerland, fixed considerable quantities of CO_2 during the winter (Keller, 1970). In December when insolation is weakest, the photosynthetic rate was still able to reach 25% of the rate in October (cf. Keller, 1965). Schmidt-Vogt and Gross (1976) measured a positive daily CO_2-balance during winter in *Picea abies* seedlings growing in the mild climate of Freiburg, Germany, so long as root systems were not frozen. A mature spruce tree near Innsbruck, Austria, still assimilated CO_2 in December and showed a high level of activity again in early March (Fig. 34). The daily total net CO_2-uptake in December was only one third of the total on a summer's day, mainly due to the shortened daylight hours rather than a reduced photosynthetic capacity (Pisek and Tranquillini, 1954).

One way to determine any trend is to measure photosynthetic capacity under optimum conditions in the laboratory using excised branches or seedlings brought in from sites at different altitude. In most cases such an approach has given data demonstrating a decrease in photosynthetic capacity with an increase in altitude. A comparison of excised material from spruce *(Picea abies)* trees showed the photosynthetic capacity at timberline near Innsbruck to be even in summer well below the rates in trees from the valley (Fig. 26; Pisek and Winkler, 1958). When using material from natural stands, altitudinal differences may in part be genetically based. Benecke (1972) placed seedlings of the same genetic origin at three altitudes (650, 1300, and 1950 m) in the same locality and determined

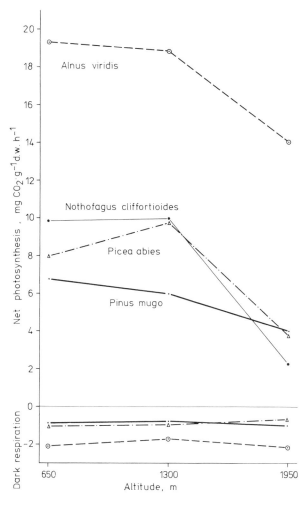

Fig. 35. Net photosynthesis and dark respiration of potted seedlings of *Alnus viridis, Nothofagus cliffortioides, Picea abies* and *Pinus mugo* grown at three elevations. Towards the end of the second growing season after establishment gas exchange was measured at 30 klux, 18° C, 7.5 mb vapour pressure deficit, 3 m s^{-1} wind and 15 C soil temperature (from Benecke, 1972)

their photosynthetic capacity in the laboratory after two growing seasons. There was no clear difference between plants at 650 and 1300 m elevation, but plants from 1950 m exhibited considerably lower photosynthetic potential (Fig. 35). This reduction from 1300 to 1950 m amounted to 25% in *Alnus viridis*, 35% in *Pinus mugo*, 61% in *Picea abies*, and 76% in *Nothofagus solandri* var. *cliffortioides*. The species known to be best adapted for growth at high altitude showed the least decline. Spruce cuttings of a number of clones grown at nearby timberline showed a similar reduction to the spruce above when compared with cuttings grown in the valley (Oberarzbacher, 1977).

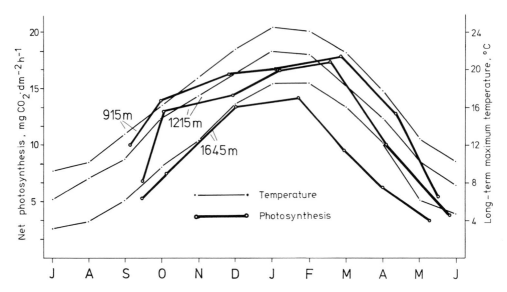

Fig. 36. Seasonal course of net photosynthesis at optimum temperature and ambient CO_2-concentration for *Eucalyptus pauciflora* at three elevations in the Snowy Mountains, Australia. Mean maximum air temperatures, with which P_N was closely correlated, are also presented (from Slatyer and Morrow, 1977)

In a field study of natural populations of an evergreen timberline tree species, *Eucalyptus pauciflora*, growing at different elevations in Australia, Slatyer and Morrow (1977) demonstrated a distinct decline in photosynthetic capacity with rising altitude. Maximum seasonal net photosynthetic rates under near natural conditions at optimum temperature declined from $18\,mg\,CO_2\,dm^{-2}h^{-1}$ at 915 m and 1214 m to $14\,mg\,CO_2\,dm^{-2}h^{-1}$ at 1645 m altitude (Fig. 36). It would seem that total seasonal CO_2-uptake at high altitude is much more reduced by the slow climb in summer of photosynthetic rates and their rapid decline in late summer than by differences in seasonal maximum rates.

Evidence of a genetic influence on reduced photosynthesis in trees at high altitude is not always clear (Neuwirth, 1969). Želawski and Góral (1966), however, in comparing high and low altitude provenances of *Pinus sylvestris* grown under identical conditions, found slightly higher photosynthetic capacities in the high-altitude provenances during spring and early summer (cf. Townsend et al., 1972).

Results of studies relating photosynthetic rates to elevation can, however, be conflicting. Mooney et al. (1964) could find no significant difference in net photosynthesis at 70 klux and 20°C of plants from various altitudes between 1350 m and 3800 m in the White Mountains, California. Dark respiration of plants from high altitude was, however, greater (cf. Maruyama et al., 1972). Though the duration of high photosynthetic capacity in *Fagus crenata* declined with altitude between 500 m and 1350 m on Mount Ninohji, Japan, the maximum rates of photosynthesis actually increased with altitude (Muruyama and Yamada, 1968).

Table 12. Chlorophyll content and assimilation quotient (gross photosynthesis: chlorophyll content) of young plants of various tree species planted at three altitudes. Chlorophyll content and gas exchange were measured under optimal conditions in the laboratory at the end of the second vegetation period after planting (from Benecke, 1972)

Altitude above sea level	mg Chlorophyll $(a+b)$ g^{-1} d.w.			Gross photosynthesis $(P_N + R_D)$ mg CO_2 mg^{-1} Chlorophyll $(a+b)$ h^{-1} ($=$assimilation quotient)		
	650 m	1,300 m	1,950 m	650 m	1,300 m	1,950 m
Pinus mugo	2.3	1.9	1.5	3.4	3.5	3.2
Picea abies	2.2	1.7	1.3	4.1	6.1	3.3
Alnus viridis	7.5	5.2	4.5	2.8	4.0	3.7

Explanation of reduced photosynthetic capacity in plants from high altitude must certainly include consideration of increased respiration (Pisek and Winkler, 1958). In some cases, especially towards the extreme upper limits, reduction in photosynthesis has also been related to reduced chlorophyll content (Table 12; Benecke, 1972). Chlorophyll seems to have been in overabundance at low altitude, especially in *Alnus viridis*, but the decline between 1300 m and 1950 m appeared to have influenced the decline in photosynthetic rate (cf. Fig. 35). For spruce seedlings at the highest altitude not only loss of chlorophyll but other radiation damage to the photosynthetic apparatus appear to have reduced assimilation performance. Between 1300 m and 1950 m the assimilation quotient (i.e., gross photosynthesis: chlorophyll content, Willstätter and Stoll, 1918) declined sharply in *Picea abies*, whereas in *Pinus mugo*, and *Alnus viridis* it changed little (Table 12).

Reduced photosynthetic capacity of twigs from timberline compared to those from low altitude could also be due to the persistent effect of previous low temperatures experienced at high altitude (Pharis et al., 1967, Neilson et al., 1972). Slatyer and Morrow (1977) obtained for *Eucalyptus pauciflora* a strong correlation between photosynthesis and mean maximum temperature of the 10-day period prior to measurement. Rook (1969) found photosynthetic capacity of *Pinus radiata* seedlings to significantly decline when they were changed from a $33°/28°$C to a $15°/10°$C day/night temperature regime. This decline was associated with a clear rise in the rate of respiration.

Whether under natural conditions at timberline the seasonal sum of photosynthesis is really less than at lower altitude can only be determined by comparative field measurements of genetically comparable material. To date, such comparative field measurements of actual seasonal net-photosynthetic sequences at different altitudes have not been published. It is thus necessary to fall back on short-term comparative measurements and calculations based on a knowledge of length of the photosynthetic period and the climatic conditions. Such an altitudinal comparison of gas-exchange in *Picea abies* and *Abies alba* was carried out by Neuwirth et al. (1966) at 1320 m, 1600 m and 2000 m in the Rila Mountains, Bulgaria, on a few summer days in August. Sun-shoots in

Table 13. Mean rates of photosynthesis (P_N) and mean radiation in the upper crown of mature spruce *(Picea abies)* at various altitudes in the Rila Mountains (Bulgaria) during August (from Neuwirth et al., 1966)

Altitude	Photosynthesis mg CO_2 g^{-1} needle d.w. h^{-1}	Radiation (cal cm^{-2} min^{-1})
1,320 m	1.02	0.78
1,600 m	1.15–0.95	0.93
2,000 m	0.81	1.16

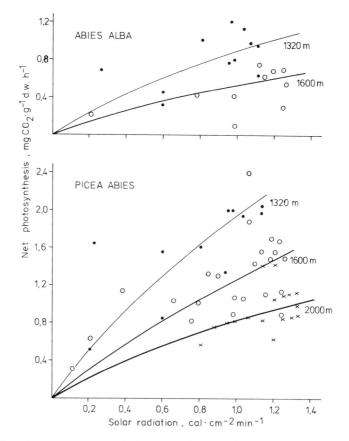

Fig. 37. Net photosynthesis in relation to radiation of shoots from the upper crown of mature *Picea abies* and *Abies alba* trees growing at different elevations in the Rila Mountains, Bulgaria. Presented data are for August and each point represents the mean of 15 measurements (from Neuwirth et al., 1966)

the upper crown assimilated about 25% less at timberline than at 1600 m and 1320 m, in spite of receiving more radiation (Table 13). At comparable levels of radiation spruce at 2000 m only showed half the photosynthetic performance of sun-shoots in a stand at 1320 m (Fig. 37).

Using measurements of photosynthetic capacity and length of photosynthetically active period Winkler (1957) calculated annual net photosynthesis for *Picea abies* in the valley as $3.9\,g\,CO_2\,g^{-1}$ d.w. a^{-1} and at timberline on Patscherkofel, Austria as only $1.8\,g\,CO_2\,g^{-1}$ d.w. a^{-1}.

From the results presented to date it can be concluded that photosynthetic primary production of trees decreases with increasing elevation above sea level predominantly because the available time for photosynthesis is drastically shortened at high altitudes. This effect is amplified by a reduction in mean photosynthetic rates at timberline.

Primary production does not always decline up the slope profile at a steady rate, due to occurrence of more favourable mid-slope temperatures which can extend the frost-free period and thereby the photosynthetically usable time by as much as 20–30 days compared to the valley floor. Above the warm mid-slope zone temperature lapse rate increases suddenly, and this contributes to dry matter production of trees in the highest belt of forest declining rapidly (Aulitzky, 1968).

3.2 Dark Respiration with Respect to Altitude and Timberline

Of all external factors, temperature is the most important in relation to dark respiration. It increases exponentially up to the highest leaf temperature of 40° C measured at timberline (Tranquillini, 1954). Minimum temperature for leaf respiration in conifers lies several degrees lower than for net photosynthesis (cf. Chap. 3.1.1.2). Keller (1965) found a minimum for dark respiration in *Pinus sylvestris*, *Abies alba*, and *Pseudotsuga menziesii* at about -8° C. It is difficult to determine exactly because respiration rates are extremely small at low temperatures.

Pisek and Winkler (1958) found reduced respiratory rates of spruce *(Picea abies)* and cembran pine *(Pinus cembra)* during the coldest time of year. In winter sun-shoots of spruce from high altitude respired between $0–5^\circ$ C at half the summer rate. This is further evidence of evergreen conifers at timberline entering true winter dormancy (Fig. 38).

Unlike the photosynthetic capacity, respiration can quickly become active at warm temperatures during winter dormancy. Respiration rates rise within a few hours of warm conditions and then slowly drop back. The stronger the preceding frost, the greater the subsequent rise in respiration (Pisek and Kemnitzer, 1968). Respiratory activity of *Abies alba* was also found to increase temporarily after a "cold shock", reaching a maximum 4–8 h after thawing, followed by a decline to normal rates within a few days. The respiratory increase in winter of frost-hardened material was much less (250% of "normal" respiration) than for frost-sensitive material in winter or summer (500% increase; Bauer et al., 1969).

This respiration response to temperature leads to considerable dry matter loss by trees during warm periods in winter. Photorespiration, though, appears to increase less than dark respiration as demonstrated in *Pinus sylvestris* (Želawski and Kucharska, 1967).

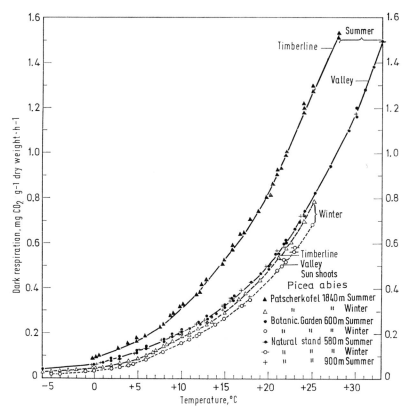

Fig. 38. Dark respiration of sun shoots taken from mature spruce *(Picea abies)* trees at low elevation (600 m) and timberline (1840 m). Measurements were carried out in summer (August) and winter (February). At both altitudes spruce showed higher rates of respiration in summer than in winter; it respired more at the timberline than in the valley (from Pisek and Winkler, 1958)

During a severe winter in the White Mountains, California, carbohydrate reserves of *Pinus aristata* trees reached exhaustion after an extensive period of negative CO_2-balance. Towards the end of this period respiration rates declined (Schulze et al., 1967).

The question whether trees have higher respiration rates at timberline than at lower elevations is of considerable importance to the seasonal CO_2-balance. Pisek and Winkler (1958) found respiration rates of high-altitude spruce *(Picea abies)* to be greater at all times and all temperatures than in the valley, especially in summer, when the difference was as much as 50% (Fig. 38, cf. Pisek and Knapp, 1959). Maruyama et al. (1972) found various broad-leaved trees from high elevations also respired more than trees from the lowland.

There is much to suggest that increased respiration at timberline is not genetically based and is solely induced by low temperatures. According to Mooney et al. (1964), a number of tree species in the White Mountains, California, possessed higher respiration rates at 20° C the higher the altitude of their origin.

Table 14. Respiration (mg CO_2 d.w.$^{-2}$ h^{-1}) of sun shoots of evergreen trees from different climatic regions (from Larcher, 1961)

Species	Mean night temperature during summer (°C)	Respiration	
		At 20° C	At mean night temperature
Stelechocarpus burahol	} ca. 25.0	0.2	0.35
Calophyllum inophyllum		0.35	0.50
Quercus glauca	24.9	0.40	0.50
Olea europaea	20.0	0.42	0.42
Quercus ilex		0.30	0.30
Picea excelsa 600 m	14.8	0.54	0.36
Picea excelsa 1,800 m	8.6	0.86	0.30

These differences disappeared, however, when plants were cultivated under identical conditions (cf. Semichatowa, 1965). Respiration of different *Pinus sylvestris* races when grown for two years in similar climates also showed no significant differences (Želawski and Góral, 1966). Against this, respiration of *Pinus radiata* seedlings increased at all test temperatures within a few days of transference from a 33°/28° C to a 15°/10° C temperature regime (Rook, 1969). *Pinus aristata* seedlings yielded a similar response when raised in the warm climate of Los Angeles where their respiration rates were only a quarter of those measured on natural plants from timberline (Mooney et al., 1966).

Increase in respiration rate resulting from cold temperatures could be a physiological adaptation thereby allowing adequate energy gain under unfavourable temperature conditions such as exist at timberline. Larcher (1961) found that plants growing in very diverse climates had similar respiration rates when measured at the mean night temperature of each site. This indicates that at similar temperature, respiration clearly increases from warm to colder regions (Table 14). This may be a reason why trees at the upper limit for their existence cannot improve their unfavourable carbon balance by restricting respiration, as do plants in conditions of extreme light deficiency but with adequate heat (Lieth, 1960).

High rates of respiration in timberline trees are detrimental to the CO_2-balance. According to Pisek and Winkler (1958) the economy coefficient of gross photosynthesis: respiration (i.e. $P_N + R_D : R_D$) was 9.3–10.4 at 12° C and 10 klux for spruce *(Picea abies)* from the valley, and 5.2–5.6 for spruce from timberline. Valley spruce thus fix ca. 10 times more than they respire whereas timberline spruce fix only ca. 5½ times more than is respired. *Pinus cembra* from timberline with a coefficient of 6.7–7.3 respired a little more economically than spruce. At 30 klux illuminance and 20° C, total spruce photosynthesis at timberline amounted to only 4 times the respiration at timberline and 6 times the respiration in the valley. By comparison sun flowers fixed 30 times the amount lost through respiration (Pisek and Winkler, 1959).

The unfavourable ratio CO_2 assimilation : respiration determined for timberline in fact influences dry matter production under natural conditions less than might be expected. At timberline cool temperatures keep respiration rates relatively low, whereas net photosynthesis attains optimum rates at quite low temperatures. Thus for the whole growing season the actual mean economy coefficient for seedlings of *Pinus cembra* was 5.1 at timberline (Tranquillini, 1959), and ranged in young trees at low elevation from 4.7 in *Fagus sylvatica* to 2.2 in *Pinus sylvestris* (Polster, 1950).

There is no reason to suggest, however, that only species with low specific respiration rates are able to advance up to high timberlines. A review by Pisek and Knapp (1959) of respiratory activity in many species shows that though *Pinus cembra*, *Pinus mugo*, and *Picea abies* belong to the species with weakest respiration rates (0.87–1.15 mg CO_2 g^{-1} d.w. h^{-1} at 20 C), *Larix decidua* from timberline respires more strongly at 2.2 mg CO_2 g^{-1} d.w. h^{-1} than *Betula*, *Fagus*, and *Quercus* in the lowland (1.5–1.7 mg CO_2 g^{-1} d.w. h^{-1}).

3.3 Carbon Balance of Trees at Timberline

Some of the carbon assimilated by leaves during the day is lost by respiration through the night. More of the carbon gain is, however, consumed by respiration in winter and by respiration of branches, stem, and roots during the whole year. By subtracting the sum of these losses from the sum of the seasonal net photosynthesis, one obtains the annual total CO_2-gain which is available for the increase of plant organic matter (cf. general outline Fig. 21).

3.3.1 Dark Respiration

In measuring CO_2 gas-exchange of *Pinus cembra* and *Larix decidua* at timberline in the Gurglertal, Austria, night respiration was also monitored for twigs with foliage (Tranquillini, 1959). Its sum for the growing season was 556 mg CO_2 g^{-1} d.w. and amounted to 8.8% of the total net-photosynthesis for *Pinus cembra*. *Larix decidua* respired more vigorously than cembran pine, in fact beween 10° and 30°C at almost twice the rate. The absolute total loss from night respiration of 708 mg CO_2 g^{-1} d.w. was thus higher, but relative to the net-photosynthetic gain the percentage loss of 7.7% was little different to that of *Pinus cembra* (Tranquillini, 1962).

A range of data collected by Tranquillini (1952) and Pisek and Tranquillini (1954) showed that in shoots from the lowland of *Picea abies* and *Fagus sylvatica* losses through night respiration were much larger than the figures quoted above. Losses varied from 4% after a very productive day and cool night to 43% after a very cloudy day and warm night. On average, leaves respired at night 19% of net photosynthesis. More recently Schulze (1970) accurately determined the night respiration loss on shoots of a mature beech tree in the Solling, Germany, as 15% of net photosynthesis.

By comparison to low altitudes, respiration losses during the night in trees at timberline are small (Fig. 39). The prime reason for this is the low night temperatures at high elevations.

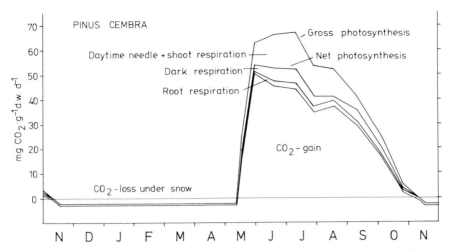

Fig. 39. Annual CO_2-balance of young *Pinus cembra* trees at timberline near Obergurgl, Austria. The seasonal course of the mean daily sum of gross photosynthesis, dark respiration of foliaged shoots measured during the day, net photosynthesis, night respiration, root respiration, and winter respiration under snow are shown. All data except root respiration and winter respiration were measured at timberline in the field. Winter CO_2-losses reduce the CO_2-gain shown for the vegetative period (from Tranquillini, 1959)

3.3.2 Winter Respiration

The sum of respiration in shoots exposed above the winter snow-cover could not be readily determined before the advent of fully climatised gas-exchange cuvettes. High insolation rates above snow formerly overheated chambers and resulted in gas-exchange no longer comparable to that under natural conditions. Against this the respiration of *Pinus cembra* seedlings below the snow-pack could be measured with adequate accuracy and without the necessity of continuous monitoring. Cembran pine seedlings at timberline in the Gurglertal, Austria, were covered for 5 months by snow and during this period they could only reassimilate a small fraction of respired CO_2. Respiration rate was, therefore, calculated from continuous temperature data of the snowed-in plants and the respiration–temperature function. For the period of snow cover the CO_2-loss amounted to $439 \, mg \, CO_2 \, g^{-1} d.w.$, i.e., 7% of the annual net-photosynthetic gain (Fig. 39).

Since CO_2-uptake at timberline is terminated during winter, plants lose weight continuously through root, needle, and shoot respiration. This dry matter loss during winter was estimated for *Pinus cembra* seedlings to be $250 \, mg \, CO_2 \, g^{-1}$ needle dry weight or one eighth of total plant weight. To recover this loss, plants needed to assimilate for 20 days after emergence from snow.

3.3.3 Root Respiration

Measurement of root respiration entails considerable methodological difficulties. Tranquillini (1959) excavated roots of *Pinus cembra* seedlings,

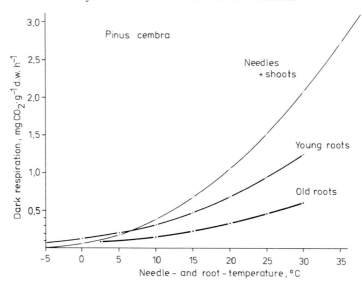

Fig. 40. Mean dark respiration of shoots and needles, thin and thick roots of *Pinus cembra* seedlings as a function of temperature. Root respiration was determined on excised material in a humid air-stream (from Tranquillini, 1959)

washed them, and then determined their CO_2 efflux in a humid air-stream at temperatures between 0–30°C. On a dry weight basis roots were found to respire at a slightly lower rate than shoots (Fig. 40). The greater the root diameter, the less was the rate of respiration per unit dry weight (cf. Yoda, 1967). Using the temperature/root respiration curve (Fig. 40) and the soil temperature patterns at 1 cm and 10 cm depth, it was possible to estimate the annual seedling root respiration total as 1863 mg CO_2 g^{-1} root d.w., i.e., in a plant with a root: needle weight ratio of only 0.25 this amounted to 7.3% of total net photosynthesis (Fig. 39).

This loss at timberline through root respiration is remarkably small by comparison with plants from lower altitude sites. According to Eidmann (1962) root respiration in various conifer seedlings amounted to 40–60% of the gross production in the growing season. During the course of a year, roots of one-year-old seedlings of *Pinus densiflora*, *Cryptomeria japonica*, and *Chamaecyparis obtusa* respired 16.6–25.0%, 12.2–13.2%, and 18.3–23.8% respectively of total net photosynthesis (Negisi, 1966). Against this, the proportion of root respiration of gross production in a 1-m tall 15-year-old stand of *Abies veitchii* was only 6.0% (Kimura et al., 1968). However, this stand was at high altitude in the subalpine zone of Mt Shimagare, Japan, and respiratory loss by roots agreed well with that of *Pinus cembra* at timberline in Austria.

The exceptionally small burden that root respiration imposes on photosynthetic production of young trees and stands at timberline could rest on the following reasons: (1) a low specific rate of root respiration in timberline tree species (2) small root: shoot ratio, and (3) low root respiration rates due to low soil temperatures. One needs to examine which of these reasons applies in the field.

Table 15. Root respiration (mg CO_2 g^{-1} d.w. h^{-1}) of young conifers at 20° C. *Pinus cembra* (Tranquillini, 1959) and *Abies veitchii* (Kimura et al., 1968) roots studied were from near timberline. Roots of other species originated from plants cultivated in lowland nurseries or glasshouses

Species	Month of year												Author
	I	II	III	IV	V	VI	VII	VIII	IX	X	XI	XII	
Larix decidua				2.73	3.09	3.12	3.40	1.80					Eidmann, 1943
Picea abies				1.02	1.28	0.58	1.68	1.25		0.64			
Abies alba				0.43	0.34	0.45	0.95	0.98		0.60			
Pinus cembra (2,050 m)			0.70										Tranquillini, 1959
Abies alba							0.61			0.28			Eidmann and Schwenke, 1967
Picea abies							0.59			0.21			
Pinus sylvestris							1.37			1.22			
Larix decidua							1.94			1.06			
Picea abies					0.46		0.49	0.63					Tranquillini, unpubl.
Larix decidua				0.37					0.73				
Picea abies (500 m)						1.75	0.88					0.66	Keller, 1967
Picea abies (1,800/2,000 m)						1.75	1.42	0.72				0.59	
Larix decidua (450 m)						0.80				0.58			
Larix decidua (1,850 m)							0.65	0.43					
Pinus sylvestris (500/650 m)						0.89						0.34	
Pinus mugo (1,850 m)							0.98	0.56				0.41	
Pinus cembra (1,850 m)							0.85					0.65	
Picea abies	0.27	0.44	0.49										Eccher, 1972
Pinus cembra	0.25	0.48	0.55										
Abies alba	0.37	0.43											
Pinus mugo	0.41	0.66											
Abies veitchii (2,340 m)					0.2	0.4	0.25	0.25		0.2		0.2	Kimura et al., 1968
Pinus densiflora	0.5	0.6	0.85	1.3	0.95	0.95	0.95	0.9	0.7	0.6	0.6	0.6	Negisi, 1966
Cryptomeria japonica	0.2	0.3	0.3	0.8	1.15	0.8	0.6	0.65	0.75	0.6	0.8	0.8	
Chamaecyparis obtusa	0.5	0.4	0.6	1.3	1.5	1.2	1.0	1.0	1.2	1.0	0.9	0.8	

When comparing root respiration in different tree species, one must take into account that respiration rates change during the course of the year (Negisi, 1966). The rate rises in spring and falls during the summer to a minimum in winter. This trend has been confirmed by other workers (Table 15). These results, however, are not always strictly comparable due to differences in method of measurement and use of glasshouse material. From Table 15 we see that rate of respiration of cembran pine roots by comparison with other young conifers grown in lowland nurseries is neither particularly high nor low. Keller's results (1967) obtained in July (summer) showed a high-altitude provenance of *Picea abies* with the highest root respiration and a high-altitude provenance of *Larix decidua* with the lowest rate. It would appear that species and races with both high and low root respiration rates can occur at timberline. Further, no significant differences in specific root respiration rates were obtained between high and low altitude provenances of *Picea abies*, *Larix decidua*, and *Pinus sylvestris* when these were raised under identical conditions (Keller, 1967).

The ratio of root dry weight to needle dry weight in young cembran pines at timberline was found to be very low (0.24). In cembran pines of similar age raised at low altitude it was 0.7 (Keller, 1967; Eccher, 1972) and in other tree seedlings mostly between 0.5 to 1.5. Likewise root development in the 15-year-old highland *Abies* stand studied by Kimura et al. (1968) was weak, with a ratio of 0.58. This is low when one bears in mind that the root:foliage ratio normally increases rapidly with stand age (Ovington, 1957). Presumably the low root mass of trees at high altitude is a function of low soil temperatures limiting root development (cf. Chap. 2.4). As a result of this, photosynthetic production by trees at timberline is less encumbered by the respiration of a smaller root mass than trees at lower altitudes.

Low soil temperatures at higher elevations above sea level per se, irrespective of root mass, ensure that organic matter loss through root respiration is kept small (cf. Fig. 40). This was demonstrated in carbon-balance trials carried out in growth cabinets (Tranquillini, unpubl.). *Larix decidua* seedlings of similar size were held from April till September under 20 klux illuminance and natural photoperiod in a $20°/16°C$ diurnal temperature cycle with roots constantly at one of three temperatures. Root systems respired on average 9.5% at $20°C$ soil temperature, 7.3% at $12°C$, and 3.7% at $4°C$ of the net photosynthesis. Keller (1967) increased root respiration in high altitude spruce *(Picea abies)* from 11% of net photosynthesis at $10°C$ root temperature to 22.5% at $20°C$ and 62% at $30°C$. Low soil temperatures at timberline, like low nightly air temperatures, are a favourable factor for the CO_2-balance of trees.

3.3.4 Stem Respiration

Since the work of Möller et al. (1954) we know that bark respiration of woody shoots, branches, and stems or trunks plays an important role in the carbon balance of trees. Depending on tree age, *Fagus sylvatica* lost 16–21% of gross photosynthetic production through this stem respiration. At timberline, stem respiration could be of special significance since it was suspected of depleting

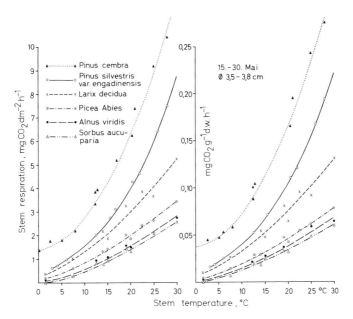

Fig. 41. Temperature dependence of stem respiration in various tree species at timberline on Patscherkofel near Innsbruck, Austria. Calculation per unit surface area (*left*) and per unit dry weight (*right*). All shoots were of 3.5–3.8 cm diameter and were collected in May (from Tranquillini and Schütz, 1970)

the limited photosynthetic production of leaves to such an extent that above a certain elevation occurrence of trees is no longer possible (Boysen Jensen, 1949).

At alpine timberline on Patscherkofel near Innsbruck stem respiration was found to vary greatly between species. Conifers such as *Pinus cembra* and *Pinus mugo* yielded the highest rates and two broadleaved species, *Alnus viridis* and *Sorbus aucuparia*, the lowest rates (Fig. 41). At the same temperature, however, stem respiration was less than for samples from low altitude (Tranquillini and Schütz, 1970).

From the temperature-dependence curves of stem respiration obtained for different times of year, and from the annual course of stem temperature on Patscherkofel, it was possible to estimate the annual cycle of total stem respiration for a mature tree of *Larix decidua* and *Pinus cembra* at timberline (Fig. 42). Stem respiration followed the stem temperature curve and attained highest values in June and July. Respiration rates were low in winter.

Total annual stem respiration in larch amounted to 4.96 kg CO_2 and was significantly less than the 10.17 kg CO_2 for cembran pine. This difference was primarily due to their different specific respiratory activity (Fig. 41), because both trees studied had the same weight of stem material. When related to gross photosynthesis, the loss through stem respiration was 18.5% in *Pinus cembra* and 12.0% in *Larix decidua*, or a loss of 23.1% and 16.9% of net photosynthesis respectively (Table 16).

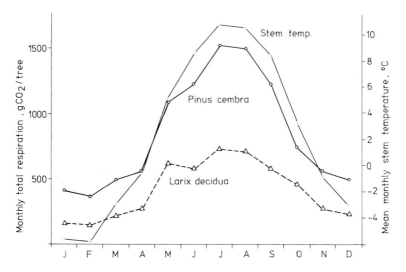

Fig. 42. Monthly sums of stem respiration in a mature tree of *Pinus cembra* and *Larix decidua*, together with the mean monthly stem temperature. *Pinus cembra*: age 76 yrs., height 7 m, diameter 20.5 cm. *Larix decidua*: age 66 yrs., height 7 m, diameter 19 cm. Respiration was estimated from the laboratory temperature dependence curves (cf. Fig. 41), 10-year monthly mean air temperatures at timberline and actual stem temperature on timberline trees (from Tranquillini and Schütz, 1970)

Table 16. Needle and shoot weight, photosynthesis and stem respiration of 7 m tall, 76-year-old trees at timberline on Patscherkofel, Austria (2,000 m) (from Tranquillini and Schütz, 1970)

	Pinus cembra	*Larix decidua*
Needle mass (g d.w.)	8,440	3,370
Annual sum of net photosynthesis (g CO_2)		
per g needle d.w.	5.2	8.7
per tree	44,023	29,319
Annual sum of gross photosynthesis (g CO_2)		
per tree	54,911	41,296
Shoot mass (g d.w.)	52,230	53,320
Annual sum of stem respiration (g CO_2)		
per tree	1,017	4,964
Stem respiration in % of:		
Gross photosynthesis	18.5	12.0
Net photosynthesis	23.1	16.9

A comparison of these results with other carbon budgets shows agreement that stem respiration reduces gross photosynthesis at timberline much less than in the valley or than in warmer regions where 25–40% may be lost through stem respiration (Tranquillini and Schütz, 1970).

3.4 Net Carbon Gain and Biomass Increment at Elevations up to Timberline

In the previous sections, photosynthesis, respiration and CO_2-balance have been dealt with as they affect dry matter production. It was shown how photosynthesis of trees declines with increasing altitude primarily as a result of shortening of the growing season, but also due to deteriorating climate for photosynthesis. Trees show an ability to adapt physiologically to the external conditions, but this does not suffice to prevent the decline in photosynthesis with rising elevation. However, the fact that carbon losses from respiration are also less at low temperature favours the positive side of the CO_2-balance at high altitude.

The reduction with altitude of annual CO_2-gain governs the increasing shortage of organic matter in the various growth zones of forest stands up the mountain. The fall-off in tree growth and increment as a result of this has been discussed in Chapters 1 and 2. The question now to be dealt with is how total organic matter increment of trees, i.e., the net dry matter production, changes with elevation and how biomass increment varies after deduction of all forms of losses (see balance sheet, Fig. 21). During the early phases of a tree's life cycle an adequate annual biomass increment is essential for its development. As trees reach maturity this increment can become infinitesimally small as long as a certain minimum net production is achieved. It is important to discover whether young trees at timberline are still able to produce a biomass increment, and whether the upper limit for existence of trees is influenced by low dry matter production limiting tree development.

3.4.1 Reduction in Dry Matter Production from Valley Floor to Timberline

The climatically induced reduction in dry matter production can readily be determined from an analysis of potted plants artifically established at various altitudes. The dry matter production in altitudinal series of plots up to timberline has been published for young seedlings of a range of tree species (e.g., Wardle, J., 1970; Wardle, 1972; Benecke and Morris, 1978). In Austria, seedling dry weight production at 1950 m (timberline) compared to 650 m (valley) was reduced by 42% in *Pinus mugo*, 54% in *Picea abies*, and 73% in *Nothofagus solandri* var. *cliffortioides* (Benecke, 1972). This dry matter decline was matched by a similar proportional decline in photosynthetic capacity (cf. Fig. 35).

The determination of net dry matter increment in mature trees and stands requires considerable effort (Newbould, 1967). Thus exact biomass data for stands at various elevations are rare. Considerably more but simpler information of timber increment is available from general applied forestry.

Annual timber volume increment of 75-year-old stands of *Picea abies* in the Central Massif, France, at 1000–1300 m elevation were 11–12 $m^3 ha^{-1}$. Above this there was a rapid fall-off in volume increment so that at the 1650 m timberline it was reduced to 1.3 $m^3 ha^{-1}$ (Oswald, 1969).

During the course of increment and yield assessments of forests up to timberline in Tyrol, Austria, the annual stand volume increment was calculated to a 6-cm top diameter (Mair, 1967). In fully stocked mixed stands of spruce–larch–cembran pine *(Picea abies–Larix decidua–Pinus cembra)* between 1600–2020 m elevation this annual volume increment was 2.0–2.5 m³ ha⁻¹ o.b. (i.e., over bark). Occasionally increments of up to 3.5 m³ ha⁻¹ were measured. In understocked and open-crowned stands which were often formerly forest pasture, volume increment dropped to 0.6–1.0 m³ ha⁻¹. By comparison, the timber volume increment of lowland stands in Tyrol lies between 4 and 6 m³ ha⁻¹ and on particularly favourable sites with a high proportion of *Abies alba* the increment can even exceed 8 m³ ha⁻¹.

Fully stocked mountain stands of *Nothofagus solandri* in New Zealand were calculated to produce an annual main stem volume increment o.b. of 6.6 m³ ha⁻¹

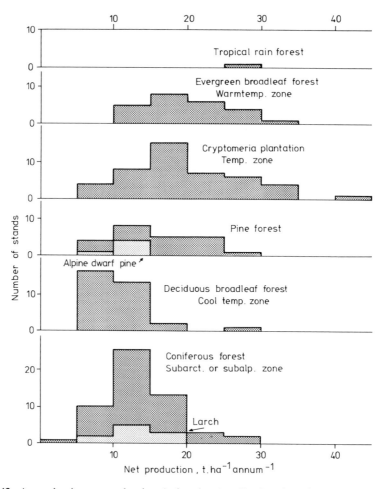

Fig. 43. Annual primary production (t d.w. ha⁻¹ a⁻¹) of various forest types in different climatic zones of the western Pacific region (from Kira and Shidei, 1967)

at 900 m and 4.5 m³ha⁻¹ at the 1340 m timberline (Wardle, J., 1970). A corresponding estimate of total annual above-ground dry matter production (i.e., including foliage and branches) gave approximate figures of 7 t ha⁻¹ at moderate elevation (900 m) and 5 t ha⁻¹ at timberline (1340 m).

A good summary of net production in various forest stands of the western Pacific region has been produced by Kira and Shidei (1967). Conifer forests in the subalpine zone and dwarf pine stands *(Pinus pumila)* in the alpine zone of the Japanese highlands in the main produced 10–15 t ha⁻¹ annually. By comparison with productivity of forests in climatically much more favourable regions this is a surprisingly high figure (Fig. 43).

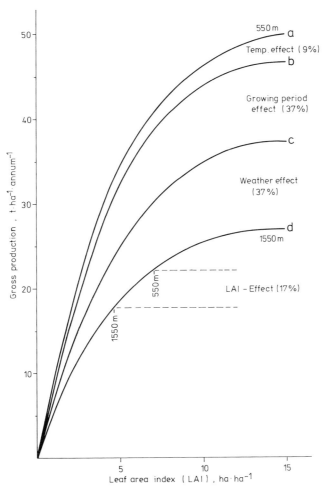

Fig. 44. Gross primary production of *Fagus crenata* on the Naeba Mountains, Japan, as a function of leaf area index (*LAI*) at 550 m elevation (*curve a*). The reduction in production at 1550 m was ascribed to individual factors: *b*, temperature at 1550 m; *c*, temperature and length of growing season at 1550 m; *d*, temperature, growing season and actual weather conditions at 1550 m. A further reduction in production is indicated due to the LAI decline of 6.8 to 4.5 from 550 to 1550 m (from Maruyama, 1971)

Mork (1942) estimated dry matter production of spruce *(Picea abies)* and birch *(Betula verrucosa)* at different altitudes in Norway. Annual spruce production declined from 8.1 t ha^{-1} at 80 m elevation to 4.5 t ha^{-1} at 180 m and 2.0 t ha^{-1} at 800 m. Net production of birch dropped from 5.2 t ha^{-1} at 180 m to only 1.2 t ha^{-1} at 800 m.

Maruyama (1971) determined dry matter production of *Fagus crenata* in the cool-temperate broadleaved forest zone on Mt Naeba (2145 m) in central Japan between 550 and 1550 m altitude, i.e., from the lower to the upper distribution of this species. Over this altitudinal range annual net production declined from 16 t ha^{-1} to 8 t ha^{-1} and corresponding annual gross production dropped from 40 t ha^{-1} to 20 t ha^{-1}. Only 9% of this production decline could be ascribed to a reduction in photosynthesis resulting from suboptimal temperatures during the growing season. A leaf area index shift from 7 to 5 could also only account for 17% of the production decline. By far the largest influence was the reduction in the length of the growing season from 185 to 122 days and the decline in radiation with increasing cloud and fog at high altitude. These two factors each accounted for 37% of the production decline (Fig. 44).

3.4.2 Primary Production of Trees in the Timberline Ecotone

As previously elucidated, above the closed forest in the open timberline ecotone (kampfzone) the micro-climate quickly becomes increasingly critical and tree height rapidly shortens over a narrow altitudinal belt. In order to evaluate the significance of dry matter production for occurrence of the upper tree limit it is important to measure net production and biomass increment of trees at the extreme highest outposts in the ecotone. Such measurements are unfortunately a rarity.

Oswald (1963) determined the dry weight of some young *Pinus cembra* trees in the ecotonal area, as well as in forest clearings at slightly lower altitude, and presented the data in relation to tree age (Fig. 45). Surprisingly the cembran pines at the highest altitude have similar biomass increment to young trees at 200 m lower altitude. Values for the upper part of the ecotone (krummholz zone) at 2190 m fall on the same "fitted" curve as values from the forest stands at 1980 m in spite of the much shorter height of individuals from the uppermost zone. The cembran pines form bushes which spread their considerable biomass as a layer close to the soil. Shidei (1963) noted that "bushy" pines at timberline can attain a dry matter density of up to 10 times greater (i.e., 10–15 kg m^{-3} stand space) than normal forest stands.

Primary production at these high altitudes is naturally considerably less than in valley sites or in regions with milder climates (cf. Tranquillini, 1959) yet there can be no question of dry matter increment being absent at these upper outposts.

Fromme (1963) suggested that biomass increase in young larch *(Larix decidua)* above timberline at altitudes between 1900–2100 m is dominantly dependent on soil type. On ranker-type soils larch attained a mean dry weight at age 20 years of 168 g, on sunny sites receiving ample precipitation even 270 g,

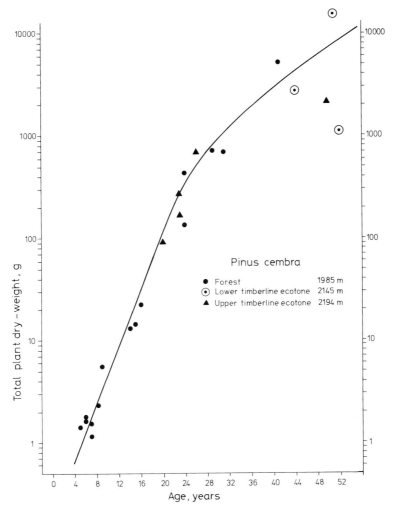

Fig. 45. Total above and below ground dry matter of *Pinus cembra* with respect to age near timberline at Obergurgl, Austria. Weight development is not noticeably different at the forest limit and in the timberline ecotone (kampfzone) (from Tranquillini, 1959; Oswald, 1963)

but the same aged trees on raw humus and moder soils achieved a dry weight of only 23 g. However, since even these latter edaphically stunted trees advance into and survive in the timberline ecotone, then the upper limit to the distribution of this species cannot be the outcome of inadequate primary production.

In the Craigieburn Range, New Zealand, Wardle (1971) found in seedling establishment trials that dry matter production of mountain beech *(Nothofagus solandri)* decreased by 60% from 1100 m to 1600 m altitude. However, even at 300 m above natural timberline (ca. 1300 m) seedlings produced more dry matter than natural seedlings within the closed stand near timberline which were subject to intense root competition. A much smaller dry matter production than that actually achieved above timberline would thus be adequate for seedling development of this species.

From these limited research results one can safely conclude that neither the rapid decrease of the tree height in the transition zone (kampfzone) above timberline, nor the total elimination of all woody plants is primarily due to inadequate dry matter production. The hypothesis that the tree limit occurs where the total photosynthetic production of leaves is consumed by respiration of the nonphotosynthetic plant organs resulting in zero net production (Boysen Jensen, 1949) is thus not valid for the alpine treeline.

3.5 Primary Production and Nitrogen Nutrition

Mineral nutrition, particularly nitrogen nutrition, plays an essential role in dry matter production of plants. An interesting question arises as to whether a deficiency in available nitrogen also contributes to the decline in primary production with increasing elevation. This supposition seems not unreasonable, since low temperatures at high altitude probably restrict microbial activity and nitrogen mineralisation in the soils.

The few studies that are relevant to this topic indicate that dry matter production at timberline is not normally limited by extreme nitrogen deficiency. Ehrhardt (1961) investigated the available nitrogen in humus of a spruce (*Picea abies*) forest at various altitudes in the Kitzbühl Alps, Austria. He found the available nitrogen in the organic soil horizon to be no less at 1880 m than at 1220 m because at the upper site the organic layer was much thicker and had a higher nitrogen content. Nitrogen concentration in spruce needles confirmed that this supply of nitrogen was in fact utilised by the timberline trees. Needle nitrogen concentrations were even higher in the highest spruce stands than in low and mid-slope stands. Ehrhardt deduced that heat as a factor limits needle and wood increment in the uppermost forest zone more than nitrogen. Thus nitrogen in the needles can be relatively speaking in over-supply.

The available nitrogen in different subalpine and alpine plant communities was studied by Rehder (1970, 1971) in the northern limestone Alps (Wetterstein Mountains, Germany). The available nitrogen was compared with dry matter production of each community. The supply of mineral nitrogen even above timberline in the alpine zone at 1800–2000 m was remarkably high. It ranged from 10 to 95 kg N ha^{-1} with an average value of 50 kg N ha^{-1}. This compares with an N-supply in Swiss broadleaved forest of 50–200 kg N ha^{-1}, in a beech forest near Göttingen, Germany, of 100 kg N ha^{-1} and in Bavarian conifer stands of 9–78 kg N ha^{-1}. In spite of the favourable amounts of available nitrogen at high altitude the communities investigated by Rehder (1971) showed annual dry matter production of only about 3 t ha^{-1} compared to 12–22 t ha^{-1} in European broadleaved and coniferous forest.

It is concluded that primary production in the alpine zone is not usually limited by insufficient available nitrogen, but dominantly by climatic factors.

4. Water Relations of Trees at Timberline

The water balance in a plant is the product of water uptake and water loss. It is primarily dependent on available water reserves in the soil and the evaporative power of the atmosphere. Changes in these two ecological factors with altitude must be known if the water relations of trees at timberline are to be understood.

Since external conditions for the water balance in trees are very different in summer and winter, and have quite different significance for trees, they will be treated separately according to season.

4.1 Water Relations in Summer

4.1.1 Precipitation and Soil Moisture

Precipitation in mountains generally increases with altitude above sea level, especially where moist air masses meet mountain ranges and are found to rise, as is the case at the northern and southern edge of the European Alps. Valleys in the lee of these outer ranges are usually relatively dry. In the various alpine regions of Switzerland and Austria total annual precipitation near timberline varies between 800 mm and 2600 mm (Turner, 1970). The inner Ötztal, Tyrol, was found to have a particularly low mean annual precipitation total of 1000 mm at timberline because it is effectively screened from the predominant rain-bearing winds (Turner, 1961b). If soil moisture deficits during summer do occur in the Alps, then this is a region where they are likely to be found.

Neuwinger (1961) determined gravimetrically the water content in the upper 10 cm of soil from June to September on sites with various aspects. Soil was decidedly drier on a shaded but wind-swept slope with NE aspect than on a sunny but wind-still slope with SW aspect, in spite of the NE site receiving more precipitation. Soil moisture seldom dropped below 20 vol%, i.e., soil water potential was always greater than -1 bar and from a physiological viewpoint plants had ample access to soil moisture.

Gunsch (1972) completed studies in the same region (Sellraintal, Tyrol) on a deforested slope (cf. Chap. 3.1.1.6). Northern aspect sites were drier with the lowest soil water potentials but even after rainless periods, which rarely last more than 14 days (Turner, 1961b), the soil potential was still > -1 bar. Exceptionally at the end of a dry spell values of down to -7 bar were obtained. Thus even in this relatively dry part of the Alps trees during summer at timberline are unlikely to experience lengthy periods with serious water uptake problems.

There are drier regions where soils desiccate even at timberline. Mooney et al. (1966) recorded the absence of precipitation in the subalpine zone of the White Mountains, California, from mid-June to the second week in August for 1963. Permanent wilting point, as defined by -15 bar soil water potential, was reached by mid-July. In spite of this, trees of *Pinus aristata* at timberline did not appear to restrict transpiration which, however, proceeded at low rates even prior to soil drought.

4.1.2 Evaporation

In spite of strong wind and intense radiation, the evaporative demand of the atmosphere is no greater at timberline than in the lowland. Tranquillini (1964b) measured evaporation of water from green Piche discs (3 cm diameter) immediately above timberline on wind-exposed Patscherkofel, Innsbruck, at 1.5 m above ground on a sunny site at 1900 m. In late summer (25 August–24 September) the highest hourly rate of evaporation was 1.08 cm^3 and the largest daily (24 h) total amounted to 10.73 cm^3. These values agreed well with earlier published results for high altitude sites. Pisek and Cartellieri (1933) recorded 1.2–1.3 cm^3 h^{-1} in the same locality and Prutzer (1961) obtained a daily total of 10.4 cm^3 for a clear day towards the end of August 1 m above various plant communities at timberline near Obergurgl, Austria.

By comparison with potential evaporation at Stuttgart (230 m) and Hohenheim (390 m), Germany (Walter, 1951), the evaporation at 2000 m in the Alps was similar, in fact, around mid-day it was even a little less. This is because the atmospheric saturation deficit is much less at high altitude due to lower temperatures (cf. Gale, 1972b).

4.1.3 Transpiration of Trees

Though difficulties in water uptake by trees at timberline during summer are unlikely to be caused by lack of soil moisture, low soil temperatures on shaded sites exposed to the wind can restrict plant water uptake. Havranek (1972) obtained a reduction in transpiration of 20% in young larch *(Larix decidua)* and spruce *(Picea abies)* when soil temperature fell from 25° to 15° C. Below a soil temperature of 15° C the transpiration was restricted even more strongly, and at 5° C plant water deficits induced stomatal closure. At the beginning and end of growing seasons plant water deficits and reduced transpiration resulting from low soil temperatures are thus possible on cool sites near timberline.

Water loss from trees at timberline corresponds to the lower evaporative power of the air at higher altitudes. Neuwirth et al. (1966) found transpiration of *Picea abies* and *Abies alba* in the Rila Mountains (Bulgaria) during August to be 49% at 1600 m and only 39% at 2000 m (timberline) compared to rates at 1320 m.

Mooney et al. (1968) compared fine weather transpiration maxima and daily totals of plants growing naturally on sites at several altitudes between 1500 and 3600 m in the White Mountains, California. At all altitudes there were

Table 17. Daily maximum transpiration rate for various tree species at timberline and in the valley on fine days during summer

Species	Month	Place and altitude	$mg\,H_2O\,g^{-1}$ $fr.w.\,h^{-1}$	$mg\,H_2O\,g^{-1}$ $d.w.\,h^{-1}$	$mg\,H_2O\,d.w.^{-2}$ h^{-1}	Author
Pinus mugo	July	Patscherkofel and Seegrube near Innsbruck 1,950 m	210	390	600	Berger-Landefeldt, 1936
Pinus cembra			280	540	660	
Picea abies			290	600	660	
Larix decidua			300	1,080	330	
Pinus sylvestris	June	Near Innsbruck 750–800 m	420		930	Pisek and Cartellieri, 1939
Picea abies			300		710	
Larix decidua			1,050		880	
Pinus sylvestris	August		300		600	
Picea abies			300		660	
Larix decidua			450		420	

species with strong and weak transpiration rates (cf. Stocker, 1956), but plants in the
subalpine and alpine zones generally used less water. Whereas transpiration rates
greater than 4000 mg g^{-1} d.w. h^{-1} and daily totals greater than 27 g g^{-1} d^{-1} were
obtained for the forest zone and the hot dry lowland, plants at and above timberline
reached corresponding figures of at most 1000 mg g^{-1} h^{-1} and 11 g g^{-1} d^{-1}.
Particularly low transpiration rates were found in the timberline pines *Pinus
flexilis* and *Pinus aristata*. At 3000 m *Pinus flexilis* transpired 400 mg g^{-1} d.w. h^{-1}
and 3.22 g g^{-1} d.w. d^{-1} as against 140 mg h^{-1} and 1.18 g d^{-1} in *Pinus aristata*.

By comparison with young trees at low altitude, timberline trees transpire
as much water or only a little less on fine days (Table 17). At timberline, however,
trees are not forced to restrict water loss (e.g., *Pinus mugo* and *Pinus cembra*)
or if so then only very temporarily (*Larix decidua* and *Picea abies*), whereas

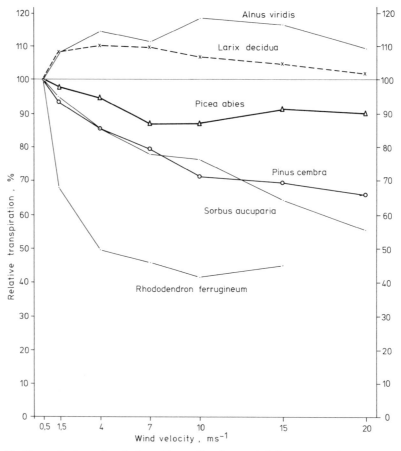

Fig. 46. Transpiration of potted seedlings of various subalpine tree species under constant
conditions of 30 klux, 20° C, 12 mb vapour pressure deficit and 15° C soil temperature
with respect to increasing wind velocity, expressed as a % of the value at 0.5 m s^{-1}. Wind
velocity was increased in steps with plants remaining at each level for 3 h. Soil in pots
was kept well watered. Each point represents two to three replicates of three to five
plants (from Tranquillini, 1969)

in the valley drastic reductions of transpiration frequently occur (Berger-Landefeldt, 1936; Pisek and Cartellieri, 1939; Pisek and Tranquillini, 1951).

The seasonal change in transpiration of *Pinus cembra* at timberline was studied in detail by Cartellieri (1935) on Patscherkofel near Innsbruck, Austria. Transpiration increased in spring only very slowly, with a rate of only $2-6\,mg\,g^{-1}$ fr. wt. h^{-1} at the beginning of April when soils had already thawed. After the end of April water loss climbed more rapidly, reaching the seasonal maximum of $200\,mg\,g^{-1}h^{-1}$ in mid-June. Transpiration rarely showed signs of reduction even at noon, a further proof that trees were adequately supplied with soil moisture at timberline.

Strong wind, to which isolated larger trees in the ecotone (kampfzone) above timberline are particularly exposed, does not significantly increase transpiration (Tranquillini, 1969). Only in *Larix decidua* and *Alnus viridis* did strong wind increase transpiration by 10–20%. In other species such as *Picea abies* and *Pinus cembra*, water loss through transpiration was clearly decreased by strong wind (Fig. 46). Natural regeneration of cembran pine in the transition zone above timberline showed no change in rate of transpiration after 24 hours of storm $(15\,m\,s^{-1})$ and only a minor increase in stomatal diffusive resistance to water vapour (Caldwell, 1970b).

4.1.4 Water Balance in Trees

Adequate soil moisture reserves, moderate evaporative demand of the high-altitude atmosphere and low specific transpiration rates of most timberline trees ensure that water balance during summer is rarely strained. Though osmotic pressures in conifers at timberline are normally higher than in most herbaceous plants at similar altitude, values do not exceed 24 bar during the growing season (Pisek et al., 1935). The timberline osmotic pressure of *Pinus cembra* cell sap between May and July was 23 bar at Patscherkofel (Cartellieri, 1935), and on one occasion exceptionally high at 27 bar (Tranquillini, 1963b).

Bearing in mind that maximum permissible osmotic pressure for cembran pine is probably close to 40 bar (Tranquillini, 1957), then all these values are relatively low and confirm the well-balanced water relations of trees at timberline.

Water potentials measured about noon on mature needles of *Picea engelmannii* and *Abies lasciocarpa* in the Medicine Bow Mountains, Wyoming, reached their seasonal peak values in July and August (Fig. 47). The maximum of -17 bar at 3300m in the timberline ecotone was only 3 bar lower than that in the forest at 2900m (Lindsay, 1971).

The water potential measured before sunrise (ψ pre-dawn)[1] in shoots of *Picea engelmannii* growing in the San Juan Mountains, Colorado, at 3500m elevation reached minimally -7 bar during the whole of the growing season (Evans, 1973).

---

[1] Maximum daily value for plant water potential is generally attained shortly before sunrise. This value corresponds approximately to the soil water potential in the root zone since plant transpiration at night declines sufficiently for an equilibrium between soil and plant water potentials to establish (Waring and Cleary, 1967).

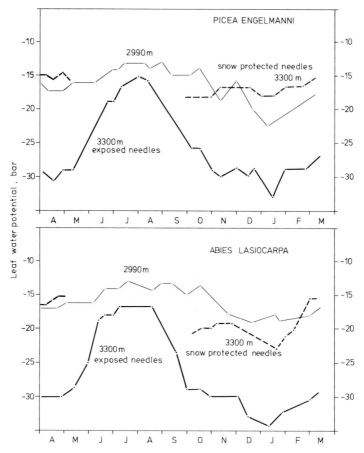

Fig. 47. Leaf water potential determined by the Schardakow method for *Picea engelmannii* and *Abies lasiocarpa* during a course of a year in the Medicine Bow Mountains, Wyoming, at 3300 m (krummholz zone) and 2990 m (near forest limit). Samples taken from below the snow during winter at the highest site showed well-balanced water relations (from Lindsay, 1971)

Water content of needles during summer on trees at timberline always remained above the critical value at which the first signs of desiccation appear (Larcher, 1957, 1972).

4.2 Water Relations in Winter

With the approach of winter, conditions critical to the plant water balance intensify. The drop of soil temperature in autumn impedes water uptake by trees (Kozlowski, 1943) and when frost penetrates the soil down to the root horizons uptake ceases completely. Whether the soil freezes or not depends largely on snow, since even a few dm of snow-cover completely insulates the soil from the cold, as was demonstrated by continuous soil temperature records

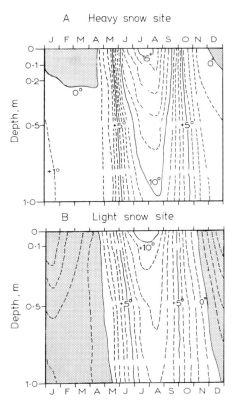

Fig. 48. Isotherms of monthly mean soil temperatures from 0 to 1 m depth on a site, **A** with heavy winter snow and **B** with little winter snow at timberline near Obergurgl, Austria (2070 m). *Hatched area* indicates when and where soil temperature falls below 0° C. The insulating effect of a snow-cover is clearly demonstrated (from Aulitzky, 1961a)

at timberline near Obergurgl, Tyrol (Aulitzky, 1961a). Under the snow-cover, soil temperature remained above −1°C even in the top 20 cm layer. At lower horizons temperatures were always above 0° C. On a wind-exposed site with little snow immediately above timberline, soil temperature was below 0° C to a depth of at least 100 cm from December to May and below −1° C from February to March. In upper soil layers temperatures of less than −4° C were even recorded (Fig. 48).

Similar results were obtained in the Front Range, Colorado (Wardle, 1968). Soil temperature barely dropped below 0° C under the canopy of the forest stand, but in the krummholz zone above timberline with little snow it was continuously below 0° C from mid-November to mid-April, and reached an extreme low of −10° C.

Turner et al. (1975) have published soil temperatures on different sites in a deforested area on Stillberg near Davos, Switzerland, at 2100–2200 m a.s.l. They were able to identify two further soil temperature types, additional to the heavy and light snow sites of Aulitzky. One is on a northerly shaded slope with much snow where in spite of long and deep snow-lie soil temperature falls below 0°C. The other is on a sunny easterly slope where in spite of little winter snow the soil freezes only intermittently for short periods to barely below 0°C.

Platter (1976) also illustrated how the snow cover affects soil temperature with measurements from the timberline ecotone on Patscherkofel near Innsbruck, Austria, in two consecutive years. The first winter was very mild but with ample snow, and soil temperature never fell in the upper 50 cm below −1°C. The following winter was colder with much less snow, and soil temperature remained below 0°C, from January to March, periodically falling to −5°C.

The water available to plants is dominantly in the medium soil pores and this freezes at −1°C (Larcher, 1957). One can thus assume that water uptake by trees on wind-exposed sites in the timberline ecotone with little or no snow cover is totally severed for 3–4 months (December–April) even in trees with roots to a soil depth of 1 m. Though trees growing on sites with a ground snow-cover generally have access to liquid water during winter, their water uptake is limited due to the low soil temperature (Kozlowski, 1943). When atmospheric conditions with strong evaporative demand exist then the water content (cf. Kleinendorst and Brouwer, 1972) and water potential (Evans, 1973) steadily decline. This situation was formerly not adequately taken into consideration as it was assumed soil frost alone was responsible for the deteriorating water balance in trees during winter at timberline.

Formation of ice in the conductive pathways of stem and branches during winter probably plays no greater role in water relations of trees at and above timberline than at lower altitudes. It is beyond dispute that during severe winters ice frequently builds up in stem and branches (Michaelis, 1934b; Larcher, 1957, 1963; Sakai, 1966, 1968; Zimmermann, 1964), thus at least temporarily halting replacement of water to the needles. However, this applies to trees in the forest stands lower down the slope to the same degree if not more so, since stems and branches are shaded, thus remaining frozen longer (Michaelis, 1934b) particularly near the base of the trunk in deep snow (Wardle, 1968).

Extreme evaporative conditions occur during late winter in the transition zone above timberline (kampfzone). High radiation levels amplified by reflection from snow-covered slopes can result in intensities up to twice those pertaining in the valley. April intensities can equal those in July (Turner, 1961a). Needle temperatures thus rise considerably above ambient air temperature in the winter sun. This "overheating" is particularly marked in late winter close to the snow surface. Data in Table 18 show an absolute needle temperature maximum of 18.4°C in March and 29.7°C in April which was close to the summer maximum. At the same time the air is still cool and has a low content of water vapour. These factors increase the leaf-air water vapour gradient which then enhances transpiration. According to Michaelis (1934a) the evaporation rate close to the snow surface during cold winter weather and high insolation rates can attain the same level as in summer on dry sites.

Trees react to the deteriorating external conditions during autumn with a step-wise reduction in water loss (Kozlowski, 1943) followed finally by complete stomatal closure. Cartellieri (1935) was able to follow the decline in transpiration in timberline cembran pine during the transition period into winter. This

Table 18. Air temperature in a screen (2 m), needle temperature of *Pinus cembra* 10 cm above ground and maximum deviation of needle temperature from ambient air temperature (°C) at timberline near Obergurgl, Austria (2,070 m) 1954/55

Month	Mean		Absolute maximum		Maximum above ambient °C	Absolute minimum		Maximum below ambient °C
	Air °C	Needles °C	Air °C	Needles °C		Air °C	Needles °C	
XI	−1.5	−3.5	11.1	13.8	4.8	−14.9	−21.2	9.3
III	−4.9	−4.9	11.0	18.4	10.5	−17.0	−19.5	4.2
IV	−1.6		10.1	29.7	21.5	−10.8	−12.0	3.8
V	3.0		14.2	30.1	18.7	− 7.2	−12.0	7.4
VI	7.3	9.0	20.6	29.5	19.3	− 4.1	− 7.5	4.4
VII	9.2	10.9	23.4	35.2	15.2	0.6	− 0.6	4.4
VIII	8.0		20.2	26.9	13.1	− 0.7	− 1.7	5.4
IX	6.1		17.2	24.6	8.9	− 6.6	− 6.9	5.8

autumnal decline and the accompanying stomatal closure may in part be initiated by low temperatures (Christersson, 1972) and endogenous reactions (Larcher, 1972). However, complete closure of stomata is certainly brought about by rapidly increasing winter water deficits. Cembran pine *(Pinus cembra)* seedlings closed their stomata at the same time as soil frost reached the root system (Tranquillini, 1957).

All published data to date indicate complete stomatal closure of timberline trees in the European Alps from November to April (Michaelis, 1934c; Cartellieri, 1935; Tranquillini and Machl-Ebner, 1971).

The further sequence of events regarding water relations of trees at timberline and in the ecotone above timberline depends very much on the snow-cover. Under snow, plants lose no water through transpiration since air within the snow is fully saturated with water vapour. Plants are able to absorb water from the wet snow and at least in late winter when infiltrating melt water thaws any frozen soil they are able to take up water through their roots. They can thus slowly alleviate any water deficits which existed before the arrival of the snow cover. At the same time osmotic pressure of cell sap falls and water content rises (Fig. 49).

Big trees are not subjected to large water deficits during the winter in spite of their crown standing above the snow surface. Small individuals appearing above snow, however, experience a distinct deterioration of the water balance up to February, osmotic pressure increasing to 30 bar and water content of needles dropping to 120% of dry matter. Comparative measurements on Patscherkofel near Innsbruck gave osmotic values up to 33.3 bar at 92% d.w. water content (Pisek et al., 1935) and up to 34.9 bar at 100% d.w. water content (Cartellieri, 1935).

That the water balance of larger trees should be better than small trees on similar sites is perhaps due to storage of water in the large xylem axes which can be drawn on to replenish needle water losses (Larcher, 1963).

Trees growing on sites in the timberline ecotone where little snow lies suffer the worst desiccation. In winter they lose 52% of autumn water reserves. Their

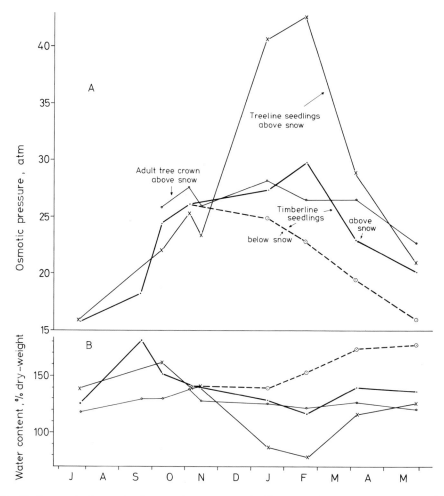

Fig. 49. Seasonal change in **A** osmotic pressure of cell sap and **B** water content (as % of d.w.) of needles from *Pinus cembra* trees at timberline near Obergurgl, Austria. Determinations were made before sunrise. The three youngest age groups of needles were pooled to form a sample. Strongest desiccation was found in young plants from sites in the timberline ecotone (kampfzone) with little winter snow (from Tranquillini, 1957)

osmotic pressure rises up to 42.5 bar and water content declines to 78% d.w. With such extreme values even *Pinus cembra* needles at their seasonal phase of highest drought resistance are subject to damage (Pisek and Larcher, 1954).

Since desiccation damage is widespread and occurs regularly in winter in the timberline ecotone (kampfzone), it will be dealt with at length in Chapter 5.5.

5. Climatic Resistance and Damage of Trees at Timberline

In the ecotone (kampfzone) above the fully stocked forest the occurrence of damage to buds, leaves, and shoots increases rapidly. Towards the upper part of the transition zone injuries become so severe that "cripple" life-forms predominate in trees, ranging from bushes with "flagged" leaders to cushions of krummholz pressed close to the ground (Fig. 50, 51). Tree form is largely dependent on the winter snow-cover, since shoots projecting above the snow are severely damaged with regularity or are completely destroyed. One can confidently state that damage which critically influences the upper limit for trees occurs predominantly during winter.

Difficulties were experienced in attempting to identify types of injury and when these occurred (Däniker, 1923). Different causes of damage produce similar symptoms which are often not immediately obvious in winter and first become visible in spring.

Better progress was made through determination of seasonal tolerance patterns in timberline trees. From such studies much is now known concerning climatic requirements of trees and the climatic limits they are just still able to tolerate. A comparison of resistance measurements with microclimatic conditions at timberline allows one to postulate which of the climatic factors become critical for trees and when damage is likely to occur.

5.1 Frost Damage

Frost resistance of timberline trees fluctuates considerably during the course of a year (Fig. 52). In summer the current seasons needles of *Pinus cembra* which are as yet not fully mature only tolerate a frost of $-2°C$ whereas 2–3 year-old needles survive frosts of $-6°C$ to $-10°C$. Frost resistance in this species increases rapidly during the autumn (September–October) and reaches a maximum in early winter (November–December) of $-40°$ to $-43°C$. Needles retain a high degree of frost resistance until April and not until May and June does frost sensitivity reappear (Schwarz, 1970). Such seasonal cycles of frost hardiness in *Pinus cembra* at timberline were described by Ulmer (1937), Pisek and Schiessl (1946), and Tranquillini (1958).

A comparison of the frost-resistance curve with air temperatures in the tree crown (Fig. 52) indicates that frost damage to mature cembran pine is unlikely to occur in winter.

Fig. 50. Flagged krummholz of *Picea abies* with a wind-deformed leader above a section free of branches and needles in the zone of severest damage just above snow surface, followed by a cushion-shaped crown close to the ground. These crown forms are characteristical for the lower part of the timberline ecotone (kampfzone)

Young *Pinus cembra* buried for 6 months under snow was found to be slightly less frost-resistant by a few degrees C (Schwarz, 1970). Larger differences were demonstrated between natural seedlings and older trees at timberline near Obergurgl, Austria, in the extremely cold winter of 1955/56 (Tranquillini, 1958). Shoots of young plants dug out of the snow were damaged by frosts 6° to 10° C higher than comparable shoots of older trees above the snow. The shoots below snow where temperatures remain close to 0° C were thus not able to develop

Fig. 51a and b. Stunted growth forms of larch (*Larix decidua*) in the Tuxer Alps at 2250 m a.s.l., Austria. Crown forms at the krummholz limit are moulded intirely by snow-lie because all twigs projecting above the protective snow-cover are regularly subject to die-back (**a**). Crowns with cornice, desk and table shapes are common (**b**)

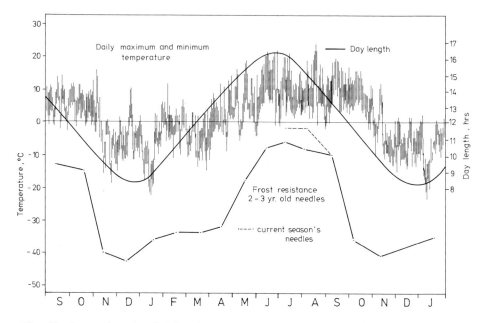

Fig. 52. Seasonal cycle of daily air max/min temperatures in a tree crown 2 m above ground, day-length and frost hardiness of needles from a mature *Pinus cembra* tree at timberline (2000 m) on Patscherkofel, Austria. The frost tolerance curve indicates the lowest temperatures at which no visible damage could be detected (from Schwarz, 1968)

the same degree of frost tolerance. During this severe winter screen air-temperature dropped to a minimum of $-30.6°$ C. Needles near the snow surface became even cooler than the air by several degrees C on clear nights (Table 18). Since young snowed-in plants of *Pinus cembra* can suffer at least partial frost damage at $-31°$ C, frost injury to this species at timberline is possible at the height of extremely cold winters, but only when young trees are suddenly blown free of snow.

Pinus cembra is the most frost-resistant tree species in the European Alps. Maximum winter frost resistance in spruce *(Picea abies)* was found to be $-36°$ to $-38°$ C and in mountain pine *(Pinus mugo)* $-35°$ C (Ulmer, 1937; Pisek and Schiessl, 1946). Thus these two species are also adequately frost-resistant to survive cold winters without harm from frost.

In mountains with even colder climates such as the Rockies in Alberta and the Taigetsu in Japan, needles of evergreen trees and shoots of deciduous species attain a frost tolerance that enables them to withstand temperatures below $-70°$ C (Sakai and Weiser, 1973). The same species, however, become frost-sensitive in summer and then suffer damage at a few degrees C below zero (Schwarz, unpubl.; Sakai and Otsuka, 1970).

At the altitude of timberline, mild frosts can occur at any time during the summer growing season. Undeveloped needles and shoots, especially those of spruce with early bud-break, are sometimes damaged by late spring frosts (Pümpel

Table 19. Comparisons of rate of temperature change when crossing the freezing range
(-4 to $-8°$ C) in *Pinus cembra* needles

	Under natural conditions at timberline			In the laboratory during determination of frost-hardiness
	Mid-winter	Late winter		
	Freezing and thawing	Freezing	Thawing	Freezing and thawing
Mean rate	$1.3°$ C h^{-1}	$5.4°$ C h^{-1}	$10.2°$ C h^{-1}	$3°$ C to $4°$ C h^{-1}
Maximum rate	$8.0°$ C h^{-1}	$2°$ C min^{-1}	$2°$ C min^{-1}	

et al., 1975). Frost damage does not threaten survival of trees in the timberline ecotone but it is a factor in determining growth deformation or krummholz (Däniker, 1923; Wardle, 1968; Holzer, 1970). Even in New Zealand's mild timberline climate, Wardle (1965) records frequent damage to new shoots from late frosts.

Severe frost damage was noted at timberline on shoots of the evergreen *Eucalyptus pauciflora* in the Snowy Mountains of SE-Australia when radiation frosts occur at night during spring (Slatyer, 1976). In this species the upper limit to its natural distribution may well be a true "frost boundary", especially since shoot desiccation during winter appears to be of no great importance.

True frost damage can also arise from exceptionally rapid freezing and thawing of foliage. It is well known that damage to artifically frozen samples is greater the more rapid their rate of cooling or thawing (Rottenburg, 1968; Pisek and Schiessl, 1946). A large part of needle water in *Pinus cembra* freezes between $-4°$ and $-8°$ C (Tranquillini and Holzer, 1958). From continuous registration of needle temperature at timberline this temperature range is known to be passed through much more rapidly at timberline in late winter than in laboratory frost tests (Table 19). This is explained by the morning sun arriving late with high intensity from behind a horizon elevated by mountains. Frozen needles are then heated rapidly and according to Holzer (1959) needles on the south (i.e., sunny) side of isolated-tree crowns in the timberline ecotone can be seriously damaged.

5.2 Ultraviolet and High Intensity Radiation Damage

A characteristic of mountain climates is their high radiation intensities. Turner (1958a) measured short-duration maxima of 2.2 cal cm^{-2} min^{-1} (equivalent to 190 klux) at timberline (2000 m) near Obergurgl, Austria, and these exceed the extraterrestrial solar constant. In winter, reflection from snow amplifies radiation up to double the intensity at low altitude. Radiation at high altitude comprises a relatively large proportion in the ultraviolet wavelengths. Short-wave radiation is more strongly scattered and absorbed than radiation of longer wavelengths in air layers at lower altitudes. The question arises whether damage from high intensity or ultraviolet radiation occurs in trees at timberline.

Cline and Salisbury (1966) studied the effect of UV-light on different plants by using a germicidal lamp (λ = 254 nm, total radiation flux density 0.076 cal cm^{-2}min^{-1}) and a high pressure Xenon lamp (continuous spectrum between 200 and 380 nm). Conifer needles proved to be remarkably UV-resistant. While many herbaceous plants died at UV levels of 1–5 cal cm^{-2}, the tolerance limits for *Picea pungens* and *Pseudotsuga menziesii* lay between 100–400 cal cm^{-2}, for *Pinus contorta* at 400–1200 cal cm^{-2}, and *Pinus nigra* as well as *Pinus ponderosa* were not damaged until UV levels reached 1200–3000 cal cm^{-2}. Under the Xenon lamp with total radiant flux density of 2 cal cm^{-2}min^{-1} including a UV-component of 0.038 cal cm^{-2}min^{-1} *Pinus nigra* needles survived for 400 h, whereas leaves of *Zea mays* showed signs of damage after only 15 h and for *Xanthium pennsylvanium* already after 3.5 h.

High UV-resistance according to Lautenschlager-Fleury (1955) rests on the filtering effect of epidermal cells. This effect increases with increasing light climate in which plants grow, and is thus particularly pronounced in alpine plants.

More recently Caldwell (1968) investigated the biologically effective medium-wave UV-radiation (280–315 nm) at altitudes between 125 m and 4350 m. Total UV-radiation for a cloudless day at 4350 m was only 26% greater than at 1670 m. Filtering out the UV-radiation with a mylar plastic sheet for 2 years produced no significant change in growth or development of the natural vegetation.

It is possible to conclude from these published results that UV-radiation does not induce damage in trees growing near timberline.

A change in colour of needles during winter is often observable on the sunny side of trees and branches of conifers. The greater the irradiance, the more intense this discolouration becomes, and it is thus more pronounced at timberline than in the lowland.

The yellowing of needles is related to a reduction in chlorophyll content (a and b), and chlorophyll was found to reach a minimum level in January and February (Tranquillini, 1957). At the same time lutein concentration was at its seasonal maximum. Lutein has a seasonal fluctuation which is the mirror image of chlorophyll while β-carotin concentration remains practically constant (Benecke, 1972).

The chlorophyll decline is governed by high radiation intensities which destroy the chlorophyll molecule photochemically (Montfort, 1950). The photo-instability of chlorophyll in conifers is particularly pronounced in winter when photosynthesis enters a "cold-induced" dormancy and energy absorbed by chlorophyll cannot be usefully converted (Lomagin and Antropova, 1966). A certain protection against strong radiation in winter is afforded by chromatophores wandering into the corners of cells *(Pinus cembra)* or concentrating around the cell nucleus (*Picea abies*; Holzer, 1958).

As long as cells of the assimilation tissue are not damaged, the loss of green pigment during winter is replenished in spring by production of new chlorophyll. In conditions of extreme light and pigment destruction, this replenishment of chlorophyll may no longer be possible, especially in the case of photosensitive plant species. The chlorophyll content in spring then remains at winter levels, i.e., well below the autumn level before winter set-in.

Benecke (1972) found that seedlings of *Alnus viridis*, *Pinus mugo*, and *Picea abies* towards the end of their second season after transplanting to various altitudes had foliar chlorophyll contents that decreased with elevation (Table 12). The reduction in chlorophyll paralleled a decline in net CO_2-uptake (Fig. 35), particularly from 1300 m to timberline. It would appear that spruce, with its marked photo-sensitivity could be lethally damaged by high radiation intensities on sunny sites in the timberline ecotone, since the photosynthetic capacity of seedlings at timberline was shown to have diminished decidedly more than could be accounted for by the decrease in chlorophyll (Benecke, 1972).

In afforestation of high-altitude sites between 2700 and 3300 m in the Rocky Mountains, Ronco (1970) found symptoms of marked chlorosis in seedlings of *Picea engelmannii* which were not explained by nitrogen or water deficiencies, but were related to strong radiation. The yellowish seedlings had low photosynthetic rates and did not survive the following winter.

The photo-instability of chlorophyll in spruce *(Picea abies)* is also apparent in investigations by Collaer (1934) in the Swiss Alps. Chlorophyll content in needles exposed to the sun declined above 1650 m elevation whereas in larch *(Larix decidua)* the reduction commenced at 1950 m and cembran pine *(Pinus cembra)* showed no change right up to timberline.

5.3 Heat Damage

The highest air temperature measured at timberline near Obergurgl, Austria, has been 27.6° C. Needles of *Pinus cembra* close to the ground can heat up to 37° C (Turner, 1958b). This is well within the limits of tolerance of cembran pine needles on seedlings and mature trees (Schwarz, 1970). Heat resistance, like frost resistance, is at a minimum in summer, but damage does not appear even in summer until the temperature climbs to 44° C (mature tree) or 46° C (young tree); (Fig. 53). These findings confirm practical experience that heat damage in needles of trees at timberline has never been observed.

Other plant organs, however, are susceptible to heat injury, particularly stems of young seedlings where these break through the soil surface, and roots or seed in the uppermost soil horizon. Soil surface above timberline can in certain circumstances temporarily become extremely warm. Turner (1958b) measured a maximum surface temperature of close to 80° C on a 35° SW slope within a raw humus area bare of vegetation (Fig. 54).

Damage to plants from heat depends on duration of the high temperature (Huber, 1935). On the above cited area with a peak of 80° C the soil surface temperature remained for 250 min above 50° C, 192 min > 60° C, 102 min > 70° C and 60 min > 75° C (Aulitzky, 1961c).

Stem temperature of *Pinus cembra* seedlings was measured on a clear day in September about noon on a hot bare site at timberline (Tranquillini and Turner, 1961). When the soil surface recorded 53.3° C the stem temperature was 41.8° C at ground level. Stem temperature thus remains significantly lower than the soil surface against which it is either well insulated (Biebl, 1962) or the transpiration stream provides internal cooling (Rouschal, 1939).

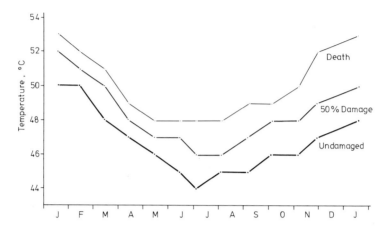

Fig. 53. Annual cycle of heat resistance in needles of a mature *Pinus cembra* tree on Patscherkofel, Austria at timberline (2000 m). Samples were placed in plastic bags and submerged in heated water for 30 min. Curves are drawn for no visible damage, 50% damage and death of all samples (from Schwarz, 1970)

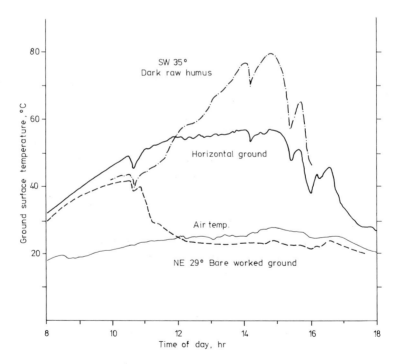

Fig. 54. Diurnal pattern of screen air temperature 2 m above ground, and soil temperature at 6 mm depth on sites with different aspects near timberline (2070 m) on an extremely warm day (from Turner, 1958 b)

Little is known concerning the heat resistance of the different tissues of the main shoot axis. Bauer (1970) determined that the cambium, bark, and wood of 4-year-old fir *(Abies alba)* seedlings were able to withstand 46°C at a time when radial growth makes these most sensitive to temperature. Roots also withstood the same temperature, but death of roots occurred after half an hour at 48°C and shoots died at 50°C. Heat damage to woody tissues of tree seedlings growing on high altitude sites subject to extreme heat is thus well within the realm of possibility (Vareschi, 1931).

The constriction of a collar of bark at ground level ("heat girdling") in conifer seedlings is a well-known phenomenon (Münch, 1913). It can occur in pine *(Pinus sylvestris)* and spruce *(Picea abies)* seedlings at a soil temperature of 46°C (Nägel, 1929) or in Douglas fir *(Pseudotsuga menziesii)* seedlings in clear-felled areas at surface temperatures of 52°–65°C resulting in death (Silen, 1960).

Heat damage to stems of conifer seedlings has also been observed in the Rocky Mountains of Alberta (Day, 1963).

By contrast dry seeds are exceptionally heat-resistant. Bartels (1958) found the germination capacity of pine *(Pinus sylvestris)* seed began to decline after an 8-h treatment at 65°C, but it was still 85% after a similar period at 85°C. Even when seed has imbibed water to saturation mortalities do not occur until temperatures rise to between 60° and 70°C (Biebl, 1962).

5.4 Mechanical Damage

5.4.1 Wind

Wind as a factor in determining the upper limit of forests was formerly given much prominence mainly because many of the crown forms in the transition zone above timberline, e.g., flagged krummholz, are clearly sculptured by wind.

Windthrow of trees and breakages of branches or crowns from wind in storms is an uncommon occurrence at timberline (Däniker, 1923). The situation is quite different in the lowland, where such damage periodically takes on catastrophic proportions. However, at timberline butt-sweep and uneven development of the crown away from the prevailing wind (flag growth Fig. 55) are clear evidence of the influence of wind pressure (Holtmeier, 1971b; Yoshino, 1973).

It is still disputed whether the numerous forms of damage to foliage and shoots exposed above the winter snow in the timberline ecotone are caused by the abrasive action of wind-borne snow and ice particles as was recently surmised by Marchand and Chabot (1978) for the krummholz of *Picea mariana* and *Abies balsamea* on Mt Washington (New Hampshire).

Holtmeier (1971b) brought evidence of ice-blast polishing the surface of tree stems in extremely windy situations (cf. Klikoff, 1965). Müller-Stoll (1954) also described necrosis of spruce *(Picea abies)* needles exposed to wind that could be explained by mechanical damage from ice-blast. However, according to Turner (1968) the earlier opinion that death of shoots of the deformation of crowns exposed above snow is due to damage from ice crystals

Fig. 55. Trees frequently produce flagged crowns on wind-exposed sites above the upper limit of closed forest stands. Crown development is much weaker on the windward-than on the lee-side. The deformation results from strongly directional wind pressure and branches lignify in the position held by the wind **(a)**. Wind flags may develop on smaller trees near the krummholz limit due to desiccation of branches on the windward-side of crowns, whereas on the lee-side vigorous branch growth is possible in the protection of snow deposits **(b)**

(Brockmann-Jerosch, 1919), can no longer be considered a valid view-point. Holzer's (1959) microscopic studies were unable to confirm damage to needle cuticles of *Pinus cembra* caused by snow polishing or whipping of branches in storms. Platter (1976), using the scanning electron-microscope, found no signs of cuticular damage to spruce *(Picea abies)* needles from a wind-exposed site at timberline on Patscherkofel near Innsbruck, Austria. Not even the sensitive outer wax layer was singificantly altered when compared with needles from the valley.

Death of shoots above the snow surface is more the result of extensive desiccation, in which wind participates largely by blowing young trees free of snow and thus exposing them to desiccation. Once dry, then shoots may secondarily be damaged by mechanical action of the wind.

5.4.2 Snow

Tree growth is not possible in avalanche tracks where snow glides down regularly. In such situations the timberline is subject to an orographic depression. Slow downward snow-creep (In der Gand, 1968) can uproot small trees and

Fig. 56. Stem deformation of *Larix decidua* on a steep slope on Patscherkofel near Innsbruck. The characteristic butt-sweep results from the alternating pressures of downward snow-creep and tree orthotropy

break off larger ones. Repeated winter snow pressure leads to a characteristic butt-sweep of the main stem in some species, e.g., larch, *Larix decidua* (Fig. 56).

These forms of snow damage occur only locally and, therefore, play no major role in the distribution of forest. Likewise damage to crowns of mature trees from the weight of deposited snow is much less common at high altitude than at mid- or low-altitude sites. Crown form at timberline is commonly well adapted to avoid snow damage with narrow pointed or columnar shapes (Fig. 16, cf. Chap. 2.1). In addition to this, the snow at high altitude is rarely wet and is thus readily blown out of crowns by strong wind.

5.5 Winter Desiccation or "Frost-drought" Damage

5.5.1 Progression of Desiccation

The most common damage to be found in the transition zone above timberline is the slow desiccation of shoots projecting above the winter snow surface. Experimental evidence for winter desiccation at timberline was first presented by Goldsmith and Smith (1926), Michaelis (1934d), Steiner (1935), and Schmidt (1936). After subsequently obtaining similar results with other tree species at timberlines the conclusion has been reached that "frost-drought", i.e., winter desiccation, is the key factor in determining treeline (Tranquillini, 1967; Wardle, 1971).

Goldsmith and Smith (1926) found the osmotic pressure of cell sap of spruce *(Picea engelmannii)* needles at optimum altitude (2700 m) in the Rocky Mountains to vary little during the year. At timberline (3750 m), however, osmotic value of needles increased during the winter, reaching a maximum value of 35 bar in April. 50 m higher at the krummholz limit osmotic pressures exceeded 50 bar, a sure indication that these needles had attained lethal water deficits.

In the Allgäu, Germany, spruce *(Picea abies)* forms timberline at 1600 m, treeline at 1750 m and the limit for krummholz lies at 1900–2000 m. Michaelis (1934d) showed that needle water content and osmotic potential differed little in spring at various elevations above sea level. Osmotic pressures lay between 24 and 30 bar. In late winter, however, a rapid decline in water content and a corresponding increase in osmotic pressure was detected. In the last uppermost outposts of krummholz record values were measured and these were related to desiccation damage (Table 20). At the same altitude (1900–2000 m) *Pinus mugo* showed no major water deficits in late winter (cf. Steiner, 1935) and in this drought-resistant species osmotic values only began to increase several hundred meters higher up the slope towards its upper limit.

A similar situation exists in the Rocky Mountains, Colorado, where at 3150 m drought-resistant species *Pinus flexilis* and *Pinus aristata* stand as erect trees

Table 20. Osmotic potential and water content of *Picea abies* needles at different altitudes in the Allgäuer Alps, Germany, during March (from Michaelis, 1934d)

Altitude (m)	Osmotic potential (at)	Water content (% d. w.)
1,480	28.8	106.1
1,670	29.2	104.8
1,820	33.1	95.1
1,940	65.7	62.6

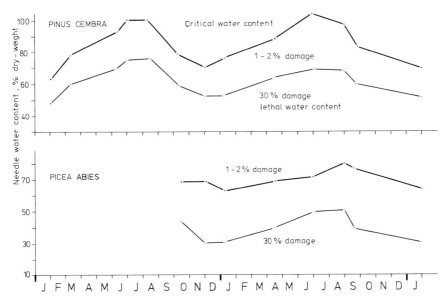

Fig. 57. Seasonal change in water content (% dry weight) of 1- and 2-year-old needles of *Pinus cembra* and *Picea abies* at various degrees of damage (from Pisek and Larcher, 1954)

above badly damaged krummholz of *Picea engelmannii* and *Abies lasiocarpa* (Wardle, 1965).

In needles of large *Picea abies* trees in the upper forest zone (1200–1500 m) at Dürrenstein, Austria, the winter osmotic pressure rose to 31 bar in March and water content fell to 113% of dry matter. In the timberline ecotone from 1700 to 1860 m foliage of stunted trees rapidly increased in osmotic pressure to a maximum of 54.7 bar with a corresponding decline in water content to 98%.

Further evidence of increasing desiccation of trees at timberline in late winter was presented by Müller-Stoll (1954) from the Black Forest, Germany. Osmotic pressure rose in March for *Picea abies* from 26.9 bar in the forest at 1240 m to 29.6 bar in the uppermost stunted spruce at 1435 m. During a dry and sunny late winter a figure of 33.1 bar was reached in April in the transition zone above timberline. Unthrifty stunted bushes with strongly wind-sculptured crowns and damaged needles reached absolute maximum osmotic values of up to 41.6 bar.

Lindsay (1971) determined the water potential of *Picea engelmannii* and *Abies lasiocarpa* needles with the Schordakow method in the forest (2290 m) and the krummholz zone (3300 m) of the Medicine Bow Mountains, Wyoming. Water potential commonly remained above −20 bar in the forest stand and in branches covered by winter snow. In branches projecting above the snow in the timberline ecotone a potential of −25 bar was reached in September, and by January this had dropped to the minimum of almost −35 bar (Fig.47).

The extent to which needles of timberline trees can withstand desiccation was investigated by Pisek and Larcher (1954). Drought resistance of *Picea abies*

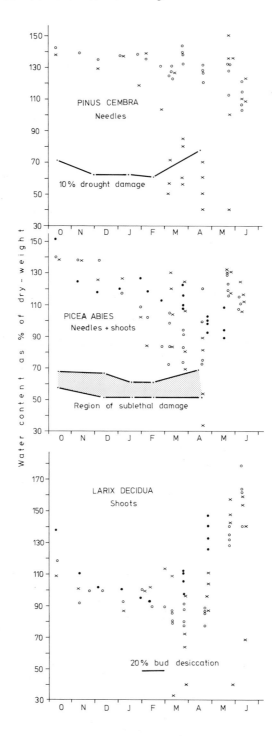

and *Pinus cembra* fluctuated during the course of the year, being greatest in winter (Fig. 57). In December the critical water content at which 1–2% of *Pinus cembra* needles are damaged was 73% of d.w. At the same moisture content spruce, *Picea abies*, begins to cast off needles. In April (late winter) *Pinus cembra* shows signs of damage when the water content sinks below 90%.

If one compares the preceding values with the water content minima found in late winter in the transition above timberline, then it becomes clear that they closely approach or even cross the limits of drought resistance (Larcher, 1957). According to Platter (1976) shoots of spruce *(Picea abies)* on Patscherkofel desiccated down to the critical "damage" level in late winter 1974/75 even at the forest limit. In the timberline ecotone above the forest the critical level was exceeded. Thus, needles of *Pinus cembra* and shoots of *Larix decidua* maintained water content just above danger levels at timberline but these species reached critical water contents in the transition area (Fig. 58).

There can be no doubt that damage to shoots and needles above the winter snow cover is caused by "frost desiccation" (Fig. 59).

5.5.2 Causes of Desiccation Damage

The climatic conditions for serious desiccation of exposed shoots on trees in the transiton zone above timberline have been dealt with in Chapter 4.2. The purpose is now to examine why frost desiccation increases so rapidly in the often quite narrow timberline ecotone of only 50–100 m when climatic factors continue to change gradually with increasing elevation. A solution to this may exist in the following two situations:

1) While the forest remains more or less closed, i.e., approaches full stocking and crown closure, the climate within the stand remains fairly uniform. Climatic maxima and minima are less extreme and snow becomes more evenly distributed. Thus winter desiccation does not reach critical levels, and within the protection of a stand individual trees can exist without damage at an altitude where they would not survive in isolation. Above a dense stand in the lower part of the timberline ecotone, climatic conditions are already so extreme that seedlings of natural regeneration are severely damaged and are unable to form a closed stand. It would only take a minor accentuation of the situation and the regeneration in the ecotone is totally destroyed.

2) The conspicuous increase in frost desiccation within the transition zone could additionally be the result of a rapid decline in drought resistance of the trees. Michaelis (1934c, d) already postulated that new shoots of trees above a certain altitude mature insufficiently before winter due to the shortness of

---

◄**Fig. 58.** Water content in needles of *Pinus cembra*, in terminal shoots and needles of *Picea abies*, and in terminal shoots after needle fall of *Larix decidua* with respect to elevation and season. Samples were from trees of natural regeneration taken in the mild winter 1974/75 from the south-side of crowns above the snow surface on the westerly slope of Patscherkofel near Innsbruck; ● valley 1000 m; ○ forest limit 1950 m; × tree limit 2100 m. For each species reference points at which first signs of damage occurred are given (cf. Fig. 57; from Platter, 1976)

Fig. 59. Severe winter desiccation of stunted *Pinus cembra* at 2150 m a.s.l. near the krummholz limit, Patscherkofel (Innsbruck). In some seasons all twigs which project above the winter snow-cover desiccate and foliage turns bright red-brown when the snow disappears during the spring

Table 21. Needle characteristics and water content of fully matured and unmatured needles of *Picea engelmannii* at timberline (from Wardle, 1968)

	Mature needles	Immature needles
Shoot length (cm)	0.7–2.7	0.4–1.2
Needle length (mm)	12–16	4–11
Needle spatial density (No. per cm shoot)	29–44	50–72
Water content (% d.w.)		
17. Nov.	134	152
22. Dec.	125	130
3. Feb.	126	51

the growing season and the consequent decline in dry matter production. The transpiration resistance of needles, i.e., cuticles and outer tissues, is then not adequate to give protection against desiccation.

This much-quoted hypothesis of Michaelis was recently restated by Wardle (1971) and used as a foundation on which to base a general theory for causes of timberline on a global scale.

The first evidence supporting the validity of this hypothesis is to be found in the work of Müller-Stoll (1954). Water content of current-year spruce *(Picea abies)* needles from the forest at 1240 m on the Feldberg, Germany, were lower in early winter than in needles from the open stand above true timberline. Shoots at lower altitude thus entered winter in more advanced state of maturation. As winter progresses, however, the situation in respect to water content reverses and needles in the timberline ecotone desiccate more strongly than in the forest stand.

According to Wardle (1968) needles of *Picea engelmannii* which fell victim to frost desiccation during the winter already possessed a pale yellowish-green colouration in the previous summer and autumn. They were also smaller and more densely spaced than needles on normally developed shoots. Their water content was higher in November (early winter) and declined during the course of the winter more rapidly than fully matured foliage (Table 21).

Holtmeier (1971a) observed in 1969 numerous cases of frost desiccation on natural regeneration of *Pinus sylvestris* at the polar timberline in Finland. The winter of observation followed a relatively cool summer in which the growing season terminated early. The needles from this cool season (1968) were not only smaller than those formed in previous seasons but were anatomically insufficiently differentiated. Amongst the features noted were a reduced number of resin canals and a thinner outer epidermal layer.

Lange and Schulze (1966) calculated that lowland spruce *(Picea abies)* requires 3 months from time of bud-burst to final thickness of the cuticular layers being formed. At high altitudes where developmental processes proceed more slowly, a longer time span is almost certainly needed. One can thus conclude that above a certain altitude, particularly in cool seasons, the growing season is too short for differentiation of new shoots to be completed and for the full cuticular protection to develop (cf. Chap. 2.1, Table 5).

5.5.3 Experimental Evidence for the Michaelis Hypothesis

Confirmation of needles and shoots from high-altitude desiccating more readily than comparable but better-developed material from lower altitude has been published (Baig et al., 1974; Tranquillini, 1974, 1976; Platter, 1976). Transpiration of excised shoots of *Pinus cembra, Picea abies*, and *Larix decidua* from various altitudes on Patscherkofel was measured in winter under constant conditions in a growth chamber. By following transpiration during desiccation it was possible to estimate cuticular transpiration. It was found that cuticular transpiration increased with increase of the elevation from which shoots were sampled (Fig. 60 a). The differences were so large that shoots from the timberline ecotone reached the critical water content at which needles begin to fall after only 75 h, whereas shoots from timberline did not reach the same stage for 130 h and in the case of shoots from the valley for 205 h (Fig. 60b).

The different desiccation sequences determined in a growth cabinet broadly matched the decline in water content of shoots at their various field sites during winter. Whereas water content at timberline always remained over 90% d.w., it reached the critical level at which desiccation damage occurs in the transition zone during February and March (Fig. 60 c). Accordingly exceptionally severe desiccation damage became visible on spruce trees in the timberline ecotone during the following spring (1973).

Similar results were obtained in comparative measurements of current season shoots of *Pinus cembra*. The differences in cuticular transpiration and rate of desiccation between timberline and transition zone were much less pronounced. It is indisputable, however, that the last outposts of cembran pine in the timberline ecotone possess less evaporative resistance and thus desiccate more quickly.

The reduction in resistance to cuticular transpiration above timberline is detectable not only in evergreen conifers. Rate of water loss from terminal shoots of the deciduous larch *(Larix decidua)* was twice as high during winter in samples from treeline as in samples from the valley (Platter, 1976).

These investigations have clearly demonstrated that the rapid increase in winter desiccation damage to trees in the timberline ecotone is not solely due to deterioration of evaporative conditions, but that above a certain elevation rapid decline in the transpiration resistance of needles also plays a vital role.

The degree of cuticular resistance to transpiration primarily depends on the length of the growing season. This resistance changes not only with altitude, but also according to the climate in different years. The longer and warmer the summer, the better are shoots able to mature, the less water is lost through the cuticle, and the less evident is desiccation damage in the following winter.

It is possible to verify the close correlation between the length of the growing season and the cuticular resistance to transpiration and consequently the winter drought resistance by artificially shortening the growing season (Tranquillini, 1974; Platter, 1976). These trials showed spruce, *Picea abies*, at timberline requires a period of 3 months of unimpaired activity in order to develop mature shoots

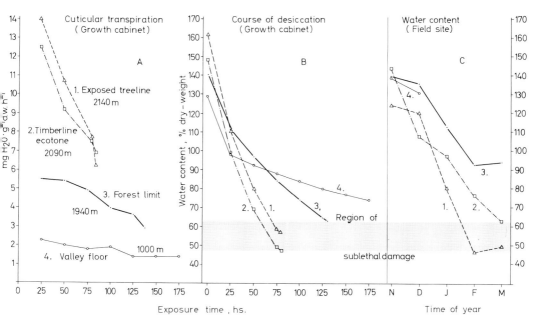

Fig. 60. A Water loss through cuticular transpiration subsequent to stomatal closure in excised current year shoots of *Picea abies* at 8–10 klux, 15° C, 8.3–9.7 mb vapour pressure deficit, 4 m s⁻¹ wind in a climatised chamber during winter (December). Needles and twigs originated from the valley (1000 m), forest limit (1940 m), timberline ecotone (2090 m) and the krummholz limit (2140 m) on Patscherkofel near Innsbruck. **B** Change in water content of shoots during progressive desiccation in a climatised chamber. Initial water content was close to the saturated value. Critical water content is *hatched*. **C** Water content of *Picea abies* shoots from trees at different altitudes on Patscherkofel during the course of winter 1972/73. Samples were collected on dry days about midday (from Baig et al., 1974)

with an effective protection against cuticular transpiration. *Pinus cembra* is less demanding than spruce in respect to length of growing season and appears to tolerate less heat for maturing needles. The pine can thus successfully cope with a shorter growing season and can advance to higher altitudes than spruce. Cembran pine is also not subject to the same degree of frost desiccation in unfavourable seasons as spruce.

5.5.4 An Analysis of Dynamic Processes at Timberline

It can be seen in the previous sections that it is the rapidly decreasing drought resistance of trees which is responsible for the increase in winter desiccation damage above timberline, and hence for the upper limit to tree growth. The deteriorating winter climate plays a subordinate role. The decline in drought resistance rests primarily on the decrease in cuticular resistance to transpiration which in turn depends on a reduction in the thickness of the cuticle. Both are closely correlated with the length of the growing season in which needles are

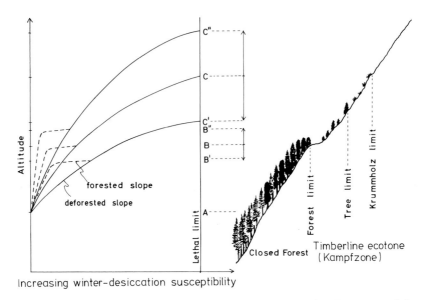

Fig. 61. Schematic representation of winter-desiccation susceptibility of trees on a deforested slope in favourable (*C″*) mean (*C*) and unfavourable (*C′*) seasons with respect to altitude. Up to elevation *B* natural regeneration can develop into a closed stand (forest limit) thereby greatly reducing the risk of winter desiccation. Above *B* (*B″* after warm seasons and *B′* after cool seasons) in the timberline ecotone susceptibility to winter desiccation rapidly increased as indicated by the broken line. The krummholz limit is determined by the resistance to desiccation of the highest tree species and fluctuates between *C′* and *C″* according to the climatic trends of seasons (from Tranquillini, 1976)

formed. Thickness and resistance of the cuticle diminish with increasing altitude; rapidly in a cool summer and more slowly in a warm summer. It now becomes clear why elevation of timberlines is so closely correlated with summer isotherms.

A similar chain of causal events can be assumed for the polar timberline where survival of trees is also determined by frost desiccation (Holtmeier, 1971a). There the picture is even clearer because winter desiccation increases with the geographical latitude, influencing exclusively the shortening of the growing season and decrease in summer heat. In lower latitude mountains such as the European Alps the desiccation effect is further amplified by the increase in atmospheric evaporation resulting from radiation conditions at high altitude.

It is now possible to explain the fluctuations in the altitudinal position of timberline during the post-glacial period after climatic change and during historical times after forest clearance (Patzelt and Bortenschlager, 1973). In Figure 61 let the risk of winter desiccation in trees begin at Altitude *A*, the risk then increases exponentially above this elevation, on the one hand due to intensification of winter conditions for desiccation and on the other due to decrease in length of growing season retarding needle development. Risk of winter desiccation progresses more slowly up the slope in favourable years with long warm summers and more rapidly in unfavourable seasons with short cool summers.

Up to altitude A (Fig. 61) trees grow unrestricted to form dense and fully stocked stands. Above this the risk of winter desiccation to regeneration spreading upwards in the zone from A to B is so slight, especially in mild seasons, that it eventually develops into fully stocked stands of erect trees, though more slowly than below elevation A. This stand then itself further reduces the occurrence of winter desiccation by ameliorating the wind climate leading to more even snow deposition which counters frost penetration of the soil. A sharp timberline is, therefore, formed at elevation B, above which the incidence of frost desiccation suddenly increases. Tree regeneration advancing further up the slope is so badly damaged above timberline that normal stands with crown closure can no longer develop. Isolated individuals or small groups do grow into trees with normal stature, but only where these occur on particularly favourable sites. The majority exist as stunted cripples. The tree deformities increase rapidly up to the krummholz limit at elevation C. This upper lethal limit lies at a different altitude for each tree species depending on their individual heat requirement for the development of cuticular protection which determines the drought resistance. In the European Alps the "lethal" limit is at lower elevation for *Picea abies* than for *Pinus cembra* and *Pinus mugo*.

Similar differences exist in the Rocky Mountains, Colorado, where the drought-resistant species *Pinus flexilis* and *Pinus aristata* tower above a severely damaged krummholz of *Picea engelmannii* and *Abies lasiocarpa* (Wardle, 1965). A further example exists in SE Europe (Rila and Pirin Mountains, Bulgaria) where the drought-resistant pines *Pinus peuce* and *Pinus leucodermis* ascend to greater altitudes than spruce, *Picea abies*.

The model presented in Figure 61 indicates that the degree of damage in the timberline ecotone (kampfzone) can vary from year to year (C'–C'') depending on the seasonal climate. A series of favourable seasons allows the tree and krummholz limit to ascend the slope. Likewise, a number of cool seasons can induce an altitudinal depression of these limits as well as the relatively stable timberline. This pattern, as described above and by the model in Figure 61, most probably accounted for the post-glacial depressions of timberline during periods of climatic deterioration. The magnitude of these depressions have been demonstrated to be about 200 m in altitude (Bortenschlager, 1977).

6. Synopsis

In mountains, the upper limit to the distribution of forests, i.e., the alpine timberline, separates as a fairly sharp discontinuity the usually closed forest from the treeless alpine heath-grassland. In principle there are three possible causes for the upper survival limit of trees on mountains with seasonally fluctuating climates, i.e., outside the tropics:
1) Negative carbon balance
2) Arrested phenological cycle
3) Inadequate resistance to deleterious factors.

6.1 Carbon Budget

The sum total net photosynthesis of tree leaves or needles during the growing season at timberline is promoted by some factors and repressed by others (Fig. 62). High light intensities, high air and soil moisture at timberline enhance photosynthesis. Genetic and physiological adaptation of photosynthesis to the timberline climate also plays a positive role. Against these, photosynthesis is restricted by low atmospheric CO_2-levels, low air and soil temperatures, high wind velocity and shortened length of growing season.

Cool nights and low air and soil temperatures check respiration losses and therefore increase the annual CO_2-gain. Long winters and high leaf temperatures resulting from intensive radiation promote respiration losses and reduce dry matter production.

Without doubt factors tending to limit dry matter production predominate at timberline, as demonstrated by the decline in dry matter production with increasing altitude. Over and above this altitudinal decline there should, however, be no additional encroachment on dry matter production in the transition zone above timberline compared to the forest below. The hypothesis that the upper survival limit for trees is determined by a zero or negative carbon balance carries no universal validity.

The reduction in dry matter production with increasing elevation does, however, lead to a deficiency in organic matter at timberline. Dry matter increment diminishes and this has a number of consequences (Fig. 63). Small shoot and leaf increment lowers photosynthetic production of the tree and this leads to a feed-back reaction. Reduction in radial growth negatively affects transport of water and assimilates as well as reducing the storage capacity of the stem

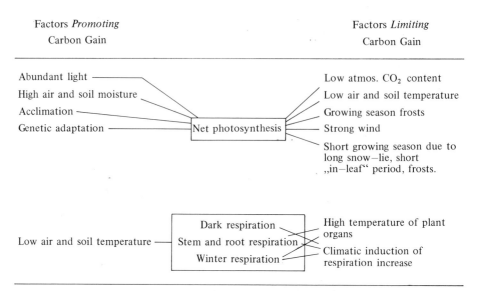

Fig. 62. Schematic diagram showing effects of various internal and external factors on net photosynthesis and dark respiration of trees at timberline with respect to dry matter production. Factors increasing respiration restrict dry matter production and factors limiting respiration increase this production

for water and organic reserves. Less root growth worsens the mineral nutrition and water supply of the tree. Reduced seed production hinders the spread of regeneration at and above timberline.

6.2 The Course of Developmental Phases

The developmental processes of interest to us here for trees can be separated into two groups:
a) seasonal periodicity in growth and developmental processes (phenology), and
b) long-term development of individual trees from seed to maturity.

Shoot extension is of special importance in the annual cycle of growth processes. It affects height growth of trees and enlargement of the leaf area. Shoot growth and its individual phases are on the one hand promoted and on the other impeded at timberline (Fig. 64). Thus growth is encouraged by a genetic disposition of leafing-out early in the season. However, the repressive influences on the developmental sequence are more numerous, as exemplified by deficient organic matter, low temperatures, and frosts during active extension, and the genetic disposition to commence bud activity late in the season.

In seasons of poor growth when the restrictive factors predominate, it can happen that above timberline the developmental cycle does not reach completion. Incomplete lignification and late formation of terminal buds enhance the chance of damage by early winter frosts.

Low total annual CO_2—uptake

Deficiency in organic matter

Reduced growth increment
of:

	Resultant effect
Shoot length and leaves	Reduction in photosynthetic productivity of the tree (positive feed—back)
Stem diameter	Smaller storage capacity for water and organic reserves, poor transport of water and assimilates
Roots	Deficient mineral nutrition and water supply
Seed	Restricted dispersal of regeneration

Fig. 63. Physiological and ecological consequences from low growth increment of trees at timberline, as a result of organic matter deficiency through weak photosynthetic productivity

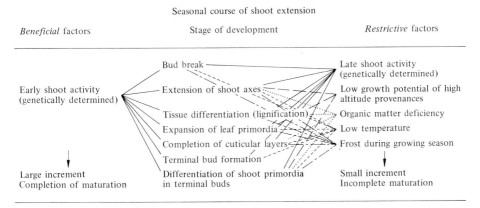

Seasonal course of shoot extension

Beneficial factors   Stage of development   *Restrictive* factors

Bud break ─────────── Late shoot activity (genetically determined)

Early shoot activity (genetically determined) ── Extension of shoot axes ── Low growth potential of high altitude provenances

Tissue differentiation (lignification) ── Organic matter deficiency

Expansion of leaf primordia ── Low temperature

Completion of cuticular layers ── Frost during growing season

Terminal bud formation

Large increment Completion of maturation   Differentiation of shoot primordia in terminal buds   Small increment Incomplete maturation

Fig. 64. Schematic interrelation of internal and external factors influencing the annual course of shoot extension in trees at timberline

Incomplete cuticular layers and leaf abscission scars increase transpiration and winter desiccation. In both these cases the phenological cycle and the associated acquisition of resistance are no longer synchronised with the climatic rhythm. Inadequate formation of shoot primordia in terminal buds results in limited shoot increment the following season, and thus in turn depresses the photosynthetic production of the tree (Fig. 65).

Recent studies (Tranquillini, 1974; Baig et al., 1974; Baig and Tranquillini, 1976; Platter, 1976) have shown that thinner cuticles and incomplete development of shoot primordia in terminal buds are widespread phenomena of tree species growing in the transition zone above timberline. These phenomena manifest

Phase of Development	Resultant Effect
	When shoot development is incomplete or not synchronised with seasonal climatic pattern
Early bud break and early leaf expansion	Late frost damage
Incomplete lignification, late terminal bud formation	Early frost damage
Incomplete development of cuticular layers and leaf abscission scars	Increased cuticular and bark transpiration ——► Winter desiccation
Inadequate differentiation of shoot primordia in the bud	Reduced growth increment ——► Decline in photosynthetic production by the tree

Fig. 65. Effects of incomplete shoot development or of shoot growth out of phase with the seasonal climatic pattern in trees at timberline

themselves after cool short summers in pronounced frost-desiccation damage the following winter, leading to particularly poor growth in the subsequent growing season.

6.3 Climatic Resistance

Many detrimental factors are of local and, therefore, overall of secondary importance to the occurrence of trees. These include heat which can kill seedlings, wind which deforms crowns on exposed sites or causes snow abrasion to damage plant organs, and parasites which periodically cause mass mortalities of trees on some sites.

There are two factors which primarily lead to injury. These are frost and desiccation. Damage of trees at timberline from both these factors rests not on any great increase in their magnitude nor on a lack of the trees ability to acquire adequate resistance, but instead on the coincident timing of these factors with growth phases at a stage of incomplete resistance.

Direct frost damage seldom occurs in the timberline ecotone because all tree tissues, even of the more frost-sensitive species, are highly frost-resistant in winter. In conifers only the new needles are initially very frost-tender and can be seriously injured at $-3°C$.

The frequency of this or lower temperatures occurring at various times of year and at three altitudes has been calculated for Patscherkofel near Innsbruck, Austria (Fig. 65a: 900 m at the lower timberline near the foot of the mountain, b: 2050 m at the upper timberline, c: 2250 m on the summit of the mountain, 100 m above the krummholz limit). The period when early and late flushing spruce extend in height, i.e., the period when needles are most frost sensitive, is shown for various years at the two lower altitudes. Spruce at 900 m begins growth early enough to be strongly susceptible to damage by late frosts. At timberline on the other hand, where frosts below $-3°C$ occur till the beginning of June, spruce commences growth so late that frost damage is quite rare. Above timberline only 200 m higher and 100 m above the local krummholz limit, no continuous period exists when plants are safe from frosts below $-3°C$. New shoots would thus be regularly frosted and this alone would prevent development

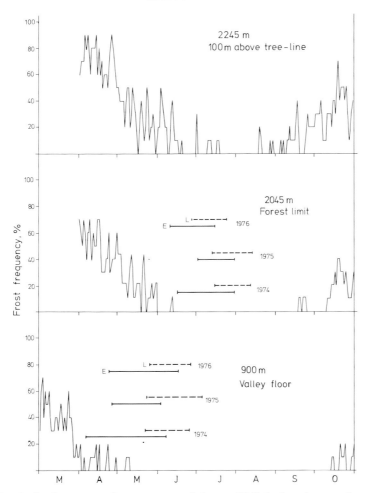

Fig. 66. Daily frequency of temperatures below $-3°C$ during the growing season for the years 1957–1966 at the weather stations Rinn, 900 m (valley floor), Patscherkofel, 2045 m (near forest limit), and Patscherkofel summit, 2245 m (100 m above krummholz limit), Austria. The period of growth in the years 1974–1976 is shown for terminal shoots of spruce (*Picea abies*) cuttings taken from *E* early- and *L* late-flushing clones grown at two elevations (from Pümpel, 1973; Lechner, 1975; Oberarzbacher, 1977; Tranquillini et al., 1978)

of trees at such altitudes. In this region, however, tree growth does not attain the upper frost limit because winter desiccation already restricts advance of trees at a slightly lower altitude. Figure 66 also clearly shows the rapid reduction with altitude of time between cessation of shoot extension and beginning of autumn frosts. The length of this period is most important for maturation of needles and above the krummholz limit there no longer exists any time for such needle development.

The major portion of damage occurring to tree species in the timberline ecotone is caused by excessive water loss from the terminal shoots during late winter. The realisation of this damage, which is decisive for the upper limit of trees, depends on a chain of cause and effect relationships (Fig. 67).

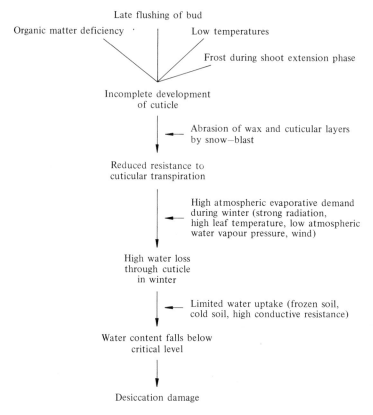

Fig. 67. Model of the causal chain of relationships leading to desiccation of terminal shoots in trees during the winter at timberline. The various environmental factors affecting this framework of interrelations are shown at their point of impact on the system

The starting point is the incomplete development of cuticular layers resulting from deficient available organic matter, low temperatures, frost during the extension growth-phase, and from a predisposition for late commencement of shoot activity. The cutinised layers and their covering of wax can additionally be reduced by abrasion when branches are in motion during storms. Consequently resistance to cuticular transpiration is reduced and the high evaporative demand of the late winter atmosphere in the transition zone above timberline causes excessive water loss through the cuticle. This transpirational water loss quickly leads to decline in foliar water content due to limited water uptake from frozen or cold soil. The critical water level at which desiccation damage begins is eventually crossed.

It has now become clear that the three possible causes for timberline, i.e., limited dry matter production, incomplete tissue maturation and inadequate climatic resistance are closely interrelated. Acting in unison they amplify the influence of each factor acting alone, and thus together ensure that above a certain altitudinal zone trees can no longer withstand the incident winter desiccation.

References

Anderson, J.E., McNaughton, S. J.: Effects of low soil temperature on transpiration photosynthesis, leaf relative water content and growth among elevationally diverse plant populations. Ecology *54*, 1220–1233 (1973)

Andre, G.S., Mooney, H.A., Wright, R.D.: The Pinyon woodland zone in the White Mountains of California. Am. Midl. Nat. *73*, 225–239 (1965)

Arno, S.F., Habeck, J.R.: Ecology of alpine Larch (*Larix lyallii* PARL.) in the Pacific Northwest. Ecol. Monogr. *42*, 417–450 (1972)

Aulitzky, H.: Die Bodentemperaturen in der Kampfzone oberhalb der Waldgrenze und im subalpinen Zirben-Lärchenwald. Mitt. Forstl. Bundesversuchsanst. Mariabrunn *59*, 153–208 (1961a)

Aulitzky, H.: Über die Windverhältnisse einer zentralalpinen Hangstation in der subalpinen Stufe. Mitt. Forstl. Bundesversuchsanst. Mariabrunn *59*, 209–230 (1961b)

Aulitzky, H.: Die Bodentemperaturverhältnisse an einer zentralalpinen Hanglage beiderseits der Waldgrenze. I. Arch. Meteorol. Geophys. Bioklimatol. B *10*, 445–532 (1961c)

Aulitzky, H.: Die Lufttemperaturverhältnisse einer zentralalpinen Hanglage. Arch. Meteorol. Geophys. Bioklimatol. Ser. B *16*, 18–69 (1968)

Baig, M.N.: Ecology of timberline vegetation in the Rocky Mountains of Alberta. Dissertation Univ. of Calgary, 1972

Baig, M.N., Tranquillini, W.: Studies on upper timberline: morphology and anatomy of Norway spruce *(Picea abies)* and stone pine *(Pinus cembra)* needles from various habitat conditions. Can. J. Bot. *54*, 1622–1632 (1976)

Baig, M.N., Tranquillini, W., Havranek, W.M.: Cuticuläre Transpiration von *Picea abies*- und *Pinus cembra*-Zweigen aus verschiedener Seehöhe und ihre Bedeutung für die winterliche Austrocknung der Bäume an der alpinen Waldgrenze. Zentralbl. Gesamte Forstwes. *91*, 195–211 (1974)

Bamberg, S., Schwarz, W., Tranquillini, W.: Influence of daylength on the photosynthetic capacity of stone pine (*Pinus cembra* L.). Ecology *48*, 264–269 (1967)

Bannan, M.W.: The vascular cambium and tree-ring development. In: Tree growth. Kozlowski, T.T. (ed.). New York: Ronald Press, 1962, pp. 3–21

Bartels, H.: Untersuchungen über die Hitzetoleranz der Koniferensamen. Forstwiss. Zentralbl. *77*, 287–294 (1958)

Bauer, H.: Hitzeresistenz und CO_2-Gaswechsel nach Hitzestress von Tanne (*Abies alba* Mill.) und Bergahorn (*Acer pseudoplatanus* L.) Dissertation Univ. Innsbruck, 1970

Bauer, H., Huter, M., Larcher, W.: Der Einfluß und die Nachwirkung von Hitze- und Kältestress auf den CO_2-Gaswechsel von Tanne und Ahorn. Ber. Dtsch. Bot. Ges. *82*, 65–70 (1969)

Bednarz, Z.: The effect of climate on the variability in the width of growth rings in stone pine (*Pinus cembra* L.) in the Tatra mountains. (Polish). Acta Agrar. et Silvestria Ser. Silvestris *16*, 17–34 (1976)

Benecke, U.: Wachstum, CO_2-Gaswechsel und Pigmentgehalt einiger Baumarten nach Ausbringung in verschiedene Höhenlagen. Angew. Bot. *46*, 117–135 (1972)

Benecke, U., Baker, G., McCracken, I.J.: Tree growth in the Craigieburn range. In: Revegetation in the rehabilitation of mountain lands. F.R.I. Symposium No. 16, N.Z. Forest Service Publication, 1978, pp. 77–98

Benecke, U., Morris, J.: Tree provenance trials. In: Revegetation in the rehabilitation of mountain lands. F.R.I. Symposium No. 16. N.Z. Forest Service Publication, 1978, pp. 99–138

Berger-Landefeldt, U.: Der Wasserhaushalt der Alpenpflanzen. Bibl. Bot. Heft *115*, Diels, L. (ed.). Stuttgart: Schweizerbartsche Verlagsbuchhandlung, 1936

Berner, L.: La limite forestriére. Feddes Repert. Berlin, Beiheft *138*, 151–161 (1959)

Biebl, R.: Protoplasmatische Ökologie der Pflanzen. Wasser und Temperatur. In: Protoplasmatologia, Band XII. Alfert, M., Bauer, H., Harding, C.V. (eds.). Wien: Springer, 1962

Bilan, V.: Effect of low temperature on root elongation in loblolly pine seedlings. XIV. IUFRO Congr. München, part 4, Sect. 23, 1967, pp. 74–82

Billings, W.D., Mooney, H.A.: The ecology of arctic and alpine plants. Biol. Rev. *43*, 481–530 (1968)

Bode, H.R.: Über den Zusammenhang zwischen Blattentfaltung und Neubildung der Saugwurzeln bei Juglans. Ber. Dtsch. Bot. Ges. *72*, 93–98 (1959)

Böhm, H.: Die Waldgrenze der Glocknergruppe. Wiss. Alpenvereinsh. (München) *21*, 143–167 (1969)

Bonnier, G.: Recherches experimentales sur l'adaptation des plantes au climat alpin. Ann. Sci. Nat., 7e Sér., Bot. *20*, 217–360 (1895)

Bonnier, G.: Nouvelles observations sur les cultures expérimentales à diverses altitudes. Rev. Gén. Bot. *32*, 305–326 (1920)

Bortenschlager, S.: Ursachen und Ausmaß postglazialer Waldgrenzschwankungen in den Ostalpen. In: Dendrochronologie und postglaziale Klimaschwankungen in Europa. Frenzel, B. (ed.). Erdwiss. Forsch. *13*, Wiesbaden: Steiner, 1977, pp. 260–266

Boysen Jensen, P.: Causal plant geography. Dansk. Vidensk. Selsk., Biol. Medd. *21*, Nr. 3 (1949)

Brockmann-Jerosch, H.: Baumgrenze und Klimacharakter. Zürich: Rascher and Cie., 1919

Caldwell, M. M.: Solar ultraviolet radiation as an ecological factor for alpine plants. Ecol. Monogr. *38*, 243–268 (1968)

Caldwell, M.M.: The effect of wind on stomatal aperture, photosynthesis and transpiration of *Rhododendron ferrugineum* L. and *Pinus cembra* L. Zentralbl. Gesamte Forstwes. *87*, 193–201 (1970a)

Caldwell, M.M.: Plant gas exchange at high wind speeds. Plant Physiol. *46*, 535–537 (1970b)

Caldwell, M.M.: The wind regime at the surface of the vegetation layer above timberline in the Central Alps. Zentralbl. Gesamte Forstwes. *87*, 65–74 (1970c)

Campell, E.: Der Tannen- oder Nußhäher und die Arvenverbreitung. Bündnerwald *4*, 3–7 (1950)

Cartellieri, E.: Jahresgang von osmotischem Wert, Transpiration und Assimilation einiger Ericaceen der alpinen Zwergstrauchheide und von *Pinus cembra*. Jahrb. Wiss. Bot. *82*, 460–506 (1935)

Christersson, L.: The transpiration rate of unhardened, hardened and dehardened seedlings of spruce and pine. Physiol. Plant. *26*, 258–263 (1972)

Cieslar, A.: Über den Ligningehalt einiger Nadelhölzer. Mitt. Forstl. Versuchswes. Österreichs *23* (1897)

Cline, M.G., Salisbury, F.B.: Effects of ultraviolett radiation on the leaves of higher plants. Radiat. Bot. *6*, 151–163 (1966)

Collaer, P.: Le rôle de la lumière dans l'établissement de la limite supérieure des forêts; observations faites dans le canton des Grisons. Ber. Schweiz. Bot. Ges. *43*, 90–125 (1934)

Däniker, A.: Biologische Studien über Baum- und Waldgrenze, insbesondere über die klimatischen Ursachen und deren Zusammenhänge. Vierteljahresschr. Naturforsch. Ges. Zürich *68*, 1–102 (1923)

Daubenmire, R.F.: Radial growth of trees at different altitudes. Bot. Gaz. *107*, 463–469 (1945)

Daubenmire, R.: Alpine timberlines in the Americas and their interpretation. Butler Univ. Bot. Stud. *11*, 119–136 (1954)

Davitaja, F.F., Mel'nik, J.S.: Radiation heating of the plants active surface and the latitudinal and altitudinal limits of forest. (Russian). Meteorol. Gidrol. Moskva 1962, 3–9 (1962)

Day, R. J.: Spruce seedling mortality caused by adverse summer microclimate in the Rocky Mountains. Dept. For. Publ. No. 1003 (1963)

Eccher, M.: Die Wurzelatmung von Holzpflanzen und ihre Bedeutung für die Stoffproduktion. Hausarbeit Bot. Inst. Innsbruck, 1972

Ehrhardt, F.: Untersuchungen über den Einfluß des Klimas auf die Stickstoffnachlieferung von Waldhumus in verschiedenen Höhenlagen der Tiroler Alpen. Forstwiss. Zentralbl. *80*, 193–215 (1961)

Eidmann, F.E.: Untersuchungen über die Wurzelatmung und Transpiration unserer Hauptholzarten. Schriftenreihe H. Göring-Akad. Dtsch. Forstwirtsch. *5* (1943)

Eidmann, F.E.: Atmung der unterirdischen Organe und Abgaben an die Mykorrhiza. In: Internationales Symposium der Baumphysiologen in Innsbruck 1961. Zusammenfassung der Vorträge und Diskussionen, 1962, pp. 43–45

Eidmann, F.E., Schwenke, H.J.: Beiträge zur Stoffproduktion, Transpiration und Wurzelatmung einiger wichtiger Baumarten. Beih. Forstwiss. Zentralbl. *23*, 7–46 (1967)

Ellenberg, H.: Leben und Kampf an den Baumgrenzen der Erde. Naturwiss. Rundsch. *19*, 133–139 (1966)

Ermich, K.: About the seasonal course of the activity of the cambium and the formation of the tree-ring of *Fagus silvatica* L. and *Abies alba* Mill. (Polish). Rocz. Dendrol. *14*, 101–109 (1960)

Evans, A.K.: Patterns of water stress in Engelmann spruce. Thesis Colorado State University Fort Collins, 1973

Fischer, F., Schmid, P., Hughes, B.R.: Anzahl und Verteilung der in der Schneedecke angesammelten Fichtensamen. Mitt. Schweiz. Anst. Forstl. Versuchswes. *35*, 459–479 (1959)

Frankhauser, F.: Der oberste Baumwuchs. Schweiz. Z. Forstwes. *52*, 1–5 (1901)

Franklin, J.F., Krueger, K.W.: Germination of true Fir and Mountain Hemlock seed on snow. J. For. *66*, 416–417 (1968)

Friedel, H.: Verlauf der alpinen Waldgrenze im Rahmen anliegender Gebirgsgelände. Mitt. Forstl. Bundesversuchsanst. Wien *75*, 81–172 (1967)

Fromme, G.: Beschreibung des Stationsgebietes in Obergurgl-Poschach. Mitt. Forstl. Bundesversuchsanst. Mariabrunn *59*, 53–65 (1961)

Fromme, G.: Über das Wachstum von Junglärchen (*Larix decidua* Mill.) auf subalpinen Standorten im Ötztal und Paznauntal (Tirol). Zentralbl. Gesamte Forstwes. *80*, 135–174 (1963)

Fryer, J.H., Ledig, F.T.: Microevolution of the photosynthetic temperature optimum in relation to the elevational complex gradient. Can. J. Bot. *50*, 1231–1235 (1972)

Fryer, J.H., Ledig, F.T., Korbobo, D.R.: Photosynthetic response of balsam fir seedlings from an altitudinal gradient. Proceed. 19th Northeast. For. Tree Improv. Conf. Orono, Maine, 1972, pp. 27–34

Furrer, E.: Kümmerfichtenbestände und Kaltluftströme in den Alpen der Ost- und Innenschweiz. Schweiz. Z. Forstwes. *117*, 720–733 (1966)

Gale, J.: Availability of carbon dioxide for photosynthesis at high altitudes. Theoretical considerations. Ecology *53*, 494–497 (1972a)

Gale, J.: Elevation and Transpiration: Some theoretical considerations with special reference to mediterranean-type climate. J. Appl. Ecol. *9*, 691–702 (1972b)

Gams, H.: Die Lunzer Kleinklimastationen und ihre Vegetation. Bioklimat. Beibl. *2*, 70–73 (1935)

Gäumann, E.: Der Einfluß der Meereshöhe auf die Dauerhaftigkeit des Lärchenholzes. Mitt. Schweiz. Anst. Forstl. Versuchswes. *25*, 327–393 (1948)

Gäumann, E., Péter-Contesse, J.: Neuere Erfahrungen über die Mistel. Schweiz. Z. Forstwes. *102*, 108–119 (1951)

Göbl, F.: Mykorrhizauntersuchungen in subalpinen Wäldern. Mitt. Forstl. Bundesversuchsanst. Wien *75*, 335–356 (1967)

Goldsmith, G.W., Smith, J.H.: Some physicochemical properties of spruce sap. Colorado College Publ. Sc. Ser. *13*, 13 (1926)

Gunsch, J.: Vergleichende ökologische Untersuchung von Kleinstandorten im Bereich der subalpinen Zirben-Waldgrenze. Diss. Univ. Innsbruck, 1972

Haller, J.R.: Variation and hybridization in ponderosa and Jeffrey pines. Univ. Calif. Publ. Bot. *34*, 123–165 (1962)

Havranek, W.M.: Über die Bedeutung der Bodentemperatur für die Photosynthese und Transpiration junger Forstpflanzen und für die Stoffproduktion an der Waldgrenze. Angew. Bot. *46*, 101–116 (1972)

Havranek, W.M., Benecke, U.: The influence of soil moisture on water potential, transpiration and photosynthesis of conifer seedlings. Plant and Soil *49*, 91–103 (1978)

Hellmers, H., Genthe, M.K., Ronco, F.: Temperature affects growth and development of Engelmann spruce. For. Sci. *16*, 447–452 (1970)

Hermes, K.: Die Lage der oberen Waldgrenze in den Gebirgen der Erde und ihr Abstand zur Schneegrenze. Kölner Geogr. Arb. Heft 5 (1955)

Hoffmann, G.: Wachstumsrhythmik der Wurzeln und Sproßachsen von Forstgehölzen. Flora *161*, 303–319 (1972)

Hoffmann, G., Lyr, H.: Charakterisierung des Wachstumsverhaltens von Pflanzen durch Wachstumsschemata. Flora *162*, 81–98 (1973)

Holtmeier, F.K.: Die ökologische Funktion des Tannenhähers im Zirben-Lärchenwald und an der Waldgrenze des Oberengadins. Journ. Ornithologie *107*, 337–345 (1966)

Holtmeier, F.K.: Zur vergleichenden Geographie und Ökologie der alpinen und polaren Waldgrenze. Dtsch. Geographentag Kiel 1969, Tagungsber. u. wiss. Abh., Wiesbaden: Steiner, 1970, pp. 519–528

Holtmeier, F.K.: Waldgrenzenstudien im nördlichen Finnisch-Lappland und angrenzenden Nordnorwegen. Rep. Kevo Subarctic Res. Stat. *8*, 53–62 (1971a)

Holtmeier, F.K.: Der Einfluß der orographischen Situation auf die Windverhältnisse im Spiegel der Vegetation. Erdkunde *25*, 178–195 (1971b)

Holtmeier, F.K.: Geoökologische Beobachtungen und Studien an der subarktischen und alpinen Waldgrenze in.vergleichender Sicht. Wiesbaden: Steiner, 1974

Holzer, K.: Die winterlichen Veränderungen der Assimilationszellen von Zirbe (*Pinus cembra* L.) und Fichte (*Picea excelsa* LINK) an der alpinen Waldgrenze. Österr. Bot.Z. *105*, 323–346 (1958)

Holzer, K.: Winterliche Schäden an Zirben nahe der alpinen Baumgrenze. Zentralbl. Gesamte Forstwes. *76*, 232–244 (1959)

Holzer, K.: Die Vererbung von physiologischen und morphologischen Eigenschaften der Fichte. I. Sämlingsuntersuchungen. Mitt. Forstl. Bundesversuchsanst. Mariabrunn *71*, 1–185 (1966)

Holzer, K.: Das Wachstum des Baumes in seiner Anpassung an zunehmende Seehöhe. Mitt. Forstl. Bundesversuchsanst. Wien *75*, 427–456 (1967)

Holzer, K.: Cold resistance in spruce. Sec. World Consultation on Forest Tree Breeding, Washington, Documents Vol. 1, 597–613 (1970)

Holzer, K.: Die Vererbung von physiologischen und morphologischen Eigenschaften der Fichte. II. Mutterbaummerkmale. Unveröffentlichtes Manuskript, 1973

Huber, B.: Der Wärmehaushalt der Pflanzen. Naturwiss. und Landwirtschaft Heft 17. Freising-München: Datterer, 1935

Hustich, I.: The scotch pine in northernmost Finland and its dependence on the climate in the last decades. Acta Bot. Fenn. *42* (1948)

In der Gand, H.: Aufforstungsversuche an einem Gleitschneehang. Ergebnisse der Winteruntersuchungen 1955/56 bis 1961/62. Mitt. Schweiz. Anst. Forstl. Versuchswes. *44*, 229–326 (1968)

Kamra, S.K., Simak, M.: Germination studies on Scots pine (*Pinus silvestris* L.) seed of different provenances under alternating and constant temperatures. Studia Forest. Suecica 62. Stockholm: Skogshögskolan, 1968

Karrasch, H.: Microclimatic studies in the Alps. Arct. Alp. Res. *5*, A55–A63 (1973)

Kaufmann, M.R., Eckard, A.N.: Water potential and temperature effects on germination of Engelmann spruce and Lodgepole pine seeds. For. Sci. *23*, 27–33 (1977)

Keller, T.: Über den winterlichen Gaswechsel der Koniferen im schweizerischen Mittelland. Schweiz. Z. Forstwes. *116*, 719–729 (1965)

Keller, T.: Beitrag zur Kenntnis der Wurzelatmung von Koniferenjungpflanzen. Proc. 14th Congr. IUFRO München IV, Sect. 23, 329–340 (1967)

Keller, T.: Über die Assimilation einer jungen Arve im Winterhalbjahr. Bündnerwald *23*, 49–54 (1970)

Kern, K.G.: Der jahreszeitliche Ablauf des Dickenwachstums von Fichten verschiedener Standorte im Trockenjahr 1959. Allg. Forst. Jagdz. *131*, 97–116 (1960)

Kimura, M., Mototani, I., Hogetsu, K.: Ecological and physiological studies on the vegetation of Mt. Shimagare. VI. Growth and dry matter production of young Abies stand. Bot. Mag. Tokyo *81*, 287–296 (1968)

Kira, T., Shidei, T.: Primary production and turnover of organic matter in different forest ecosystems of the Western Pacific. Jpn. J. Ecol. *17*, 70–87 (1967)

Kleinendorst, A., Brouwer, R.: The effect of local cooling on growth and water content of plants. Neth. J. Agric. Sci. *20*, 203–217 (1972)

Klikoff, L.G.: Microenvironmental influence on vegetational pattern near timberline in the central Sierra Nevada. Ecol. Monogr. *35*, 187–211 (1965)

Koch, W.: Untersuchungen über die Wirkung von CO_2 auf die Photosynthese einiger Holzgewächse unter Laboratoriumsbedingungen. Flora (Jena) B *158*, 402–428 (1969)

Köstler, J.N., Mayer, H.: Waldgrenzen im Berchtesgadener Land. Jahrb. Ver. Schutze Alpenpfl. Tiere (München) *35*, 1–35 (1970)

Kozlowski, T.T.: Transpiration rates of some forest tree species during the dormant season. Plant Physiol. *48*, 252–260 (1943)

Krempl, H.: Holzbildung der Fichte in verschiedenen Seehöhen. Unveröffentlichtes Manuskript, 1978

Kuoch, R.: Der Samenanfall 1962/63 an der oberen Fichtenwaldgrenze im Sertigtal. Mitt. Schweiz. Anst. Forstl. Versuchswes. *41*, 63–85 (1965)

Kuoch, R., Amiet, R.: Die Verjüngung im Bereich der oberen Waldgrenze der Alpen. Mitt. Schweiz. Anst. Forstl. Versuchswes. *46*, 159–328 (1970)

Lamarche, V.C., Fritts, H.C.: Tree rings, glacial advance and climate in the Alps. Z.Gletscherkd. Glazialgeol. *7*, 125–131 (1971)

Lamarche, V.C., Mooney, H.A.: Recent climatic change and development of the bristlecone pine (*Pinus longaeva* Bailey) krummholz zone. Mt. Washington, Nevada. Arct. Alp. Res. *4*, 61–72 (1972)

Lange, O.L., Schulze, E.D.: Untersuchungen über die Dickenentwicklung der kutikularen Zellwandschichten bei der Fichtennadel. Forstwiss. Zentralbl. *85*, 27–38 (1966)

Larcher, W.: Frosttrocknis an der Waldgrenze und in der alpinen Zwergstrauchheide. Veröff. Museum Ferdinandeum Innsbruck *37*, 49–81 (1957)

Larcher, W.: Jahresgang des Assimilations- und Respirationsvermögens von *Olea europaea* L. ssp. *sativa* HOFF. et LINK., *Quercus ilex* L. und *Quercus pubescens* WILLD. aus dem nördlichen Gardaseegebiet. Planta *56*, 575–606 (1961)

Larcher, W.: Zur spätwinterlichen Erschwerung der Wasserbilanz von Holzpflanzen an der Waldgrenze. Ber. Naturwiss. Med. Ver. Innsbruck *53*, 125–137 (1963)

Larcher, W.: Der Wasserhaushalt immergrüner Pflanzen im Winter. Ber. Dtsch. Bot. Ges. *85*, 315–327 (1972)

Lautenschlager-Fleury, D.: Über die Ultraviolettdurchlässigkeit von Blattepidermen. Ber. Schweiz. Bot. Ges. *65*, 343–386 (1955)

Lechner, F.: Austrieb und Zuwachs in verschiedenen Höhenlagen, sowie Photosynthese von unterschiedlichen Klonen der Fichte. Diss. Univ. Innsbruck, 1975

Lechner, F., Holzer, K., Tranquillini, W.: Über Austrieb und Zuwachs von Fichtenklonen in verschiedener Seehöhe. Silvae Genet. *26*, 33–41 (1977)

Lieth, H.: Über den Lichtkompensationspunkt der Landpflanzen. Planta *54*, 530–576 (1960)

Lindsay, J.H.: Annual cycle of leaf water potential in *Picea engelmannii* and *Abies lasiocarpa* at timberline in Wyoming. Arct. Alp. Res. *3*, 131–138 (1971)

Lomagin, A.G., Antropova, T.A.: Photodynamic injury to heated leaves. Planta *68*, 297–309 (1966)

Lüdi, W.: Alter, Zuwachs und Fruchbarkeit der Fichten *(Picea excelsa)* im Alpengarten Schinigeplatte. Schweiz. Z. Forstwes. *89*, 104–110 (1938)

Ludlow, M.M., Jarvis, P.G.: Photosynthesis in Sitka Spruce *(Picea sitchensis* (Bong.) Carr.). 1. General characteristics. J. Appl. Ecol. *8*, 925–953 (1971)

Lumbe, C.: Untersuchungen über optimale Anzuchtbedingungen bei Jungzirben. Forstl. Bundesversuchsanst. Wien, Informationsdienst *80* (1964)

Machl, I.: Über den Einfluß der Wärme auf das Photosynthesevermögen einiger subalpiner Baumarten und der Alpenrose während der winterlichen Ruheperiode und auf das Austreiben der Lärche. Diss. Univ. Innsbruck, 1969

Mair, N.: Zuwachs- und Ertragsleistung subalpiner Wälder. Mitt. Forstl. Bundesversuchsanst. Wien *75*, 383–426 (1967)

Marchand, P.J., Chabot, B.F.: Winter water relations of tree-line plant species on Mt. Washington, New Hampshire. Arctic and Alpine Res. *10*, 105–116 (1978)

Marek, R.: Waldgrenzenstudien in den österreichischen Alpen. Gotha: Justus Perthes, 1910

Mark, A.F., Sanderson, F.R.: The altitudinal gradient in forest composition, structure and regeneration in the Hollyford valley, Fiordland. Proc. N.Z. Ecol. Soc. *9*, 17–26 (1962)

Maruyama, K.: Effect of altitude on dry matter production of primeval japanese beech forest communities in Naeba mountains. Mem. Fac. Agric., Niigata Univ. Nr. *9*, 87–171 (1971)

Maruyama, K., Yamada, M.: Ecological studies on beech forest. 16. Seasonal course on apparent photosynthesis and respiration rate in detached leaves of Japanese beech at different altitudes. Bull. Niigata Univ. For. *3*, 17–26 (1968)

Maruyama, K., Yanagisawa, T., Kanai, C.: Ecological studies on natural beech forest. 24. Rates of photosynthesis of detached sun-leaves in some Japanese deciduous broad-leaved tree species growing at beech zone. Niigata Agric. Sci. *24*, 13–21 (1972)

Mayer, H.: Gebirgswaldbau Schutzwaldpflege. Stuttgart: Gustav Fischer, 1976

McGee, C.E.: Elevation of seed sources and planting sites affects phenology and development of red oak seedlings. Forest Sci. *20*, 160–164 (1974)

McNaughton, S.J., Campbell, R.S., Freyer, R.A., Mylroie, J.E., Rodland, K.D.: Photosynthetic properties and root chilling responses of altitudinal ecotypes of *Typha latifolia* L. Ecology 55, 168–172 (1974)

Michaelis, P.: Ökologische Studien an der alpinen Baumgrenze. II. Die Schichtung der Windgeschwindigkeit, Lufttemperatur und Evaporation über einer Schneefläche. Beih. Bot. Zentralbl. *52* B, 310–332 (1934a)

Michaelis, P.: Ökologische Studien an der alpinen Baumgrenze. III. Über die winterlichen Temperaturen der pflanzlichen Organe, insbesondere der Fichte. Beih. Bot. Zentralbl. *52* B, 333–377 (1934b)

Michaelis, P.: Ökologische Studien an der alpinen Baumgrenze. IV. Zur Kenntnis des winterlichen Wasserhaushaltes. Jahrb. Wiss. Bot. *80*, 169–298 (1934c)

Michaelis, P.: Ökologische Studien an der alpinen Baumgrenze. V. Osmotischer Wert und Wassergehalt während des Winters in den verschiedenen Höhenlagen. Jahrb. Wiss. Bot. *80*, 337–362 (1934d)

Mlinšek, D.: Möglichkeiten der Ertragssteigerung im subalpinen Fichtenwald. In: 100-Jahrfeier Hochschule für Bodenkultur Wien, Band 4, Teil 1, 51–69 (1973)

Möller, C.M., Müller, D., Nielsen, J.: Ein Diagramm der Stoffproduktion im Buchenwald. Ber. Schweiz. Bot. Ges. *64*, 487–494 (1954)

Montfort, C.: Photochemische Wirkungen des Höhenklimas auf die Chloroplasten photolabiler Pflanzen im Mittel- und Hochgebirge. Z. Naturforsch. *5* B, 221–226 (1950)

Mooney, H.A., Brayton, R., West, M.: Transpiration intensity as related to vegetation zonation in the White Mountains of California. Am. Midl. Nat. *80*, 407–412 (1968)

Moonley, H.A., Shropshire, F.: Population variability in temperature related photosynthetic acclimation. Oecol. Plant. *2*, 1–13 (1967)

Mooney, H.A., Strain, B.R., West, M.: Photosynthetic efficiency at reduced carbon dioxide tensions. Ecology *47*, 490–491 (1966)

Mooney, H.A., West, M.: Photosynthetic acclimation of plants of diverse origin. Am. J. Bot. *51*, 825–827 (1964)

Mooney, H.A., West, M., Brayton, R.: Field measurements of the metabolic responses of bristlecone pine and big sagebrush in the White Mountains of California. Bot. Gaz. *127*, 105–113 (1966)

Mooney, H.A., Wright, R.D., Strain, B.R.: The gas exchange capacity of plants in relation to vegetation zonation in the White Mountains of California. Am. Midl. Nat. *72*, 281–297 (1964)

Morgenthal, J.: Die Nadelgehölze. Stuttgart: Gustav Fischer, 1955

Mork, E.: Über den Streufall in unseren Wäldern. (Norwegian). Medd. Nor. Skogforsögsves. Oslo *29*, 297–365 (1942)

Mork, E.: On the relationship between temperature, leading shoot increment and the growth and lignification of the annual ring in Norway spruce (*Picea abies* (L.) Karst.). (Norwegian) Medd. Nors. Skogforsöksves. *56*, 229–261 (1960)

Moser, M.: Der Einfluß tiefer Temperaturen auf das Wachstum und die Lebenstätigkeit höherer Pilze mit spezieller Berücksichtigung von Mykorrhizapilzen. Sydowia, Ann. Mycol., Ser. II, *12*, 386–399 (1958)

Moser, M.: Die ektotrophe Ernährungsweise an der Waldgrenze. Mitt. Forstl. Bundesversuchsanst. Wien *75*, 357–380 (1967)

Müller-Stoll, W.R.: Beiträge zur Ökologie der Waldgrenze am Feldberg im Schwarzwald. Angew. Pflanzensoziologie. Festschrift Erwin Aichinger, Band 2. Janchen, E. (ed.). Wien: Springer, 1954, pp. 824–847

Münch, E.: Hitzeschäden an Waldpflanzen. Naturwiss. Z. Forst. Landwirtsch. *11*, 557–562 (1913)

Münch, E.: Die Knospenentfaltung der Fichte und die Spätfrostgefahr. Allg. Forst Jagdz. 1923, 241–265 (1923)

Nägel, A.: Untersuchungen über die Maximaltemperatur von Keimlingen der Kiefer und der Fichte. Diplomarbeit Tharandt, 1929

Nägeli, W.: Einfluß der Herkunft des Samens auf die Eigenschaften forstlicher Holzgewächse IV. Die Fichte. Mitt. Schweiz. Zentralanst. Forstl. Verswes. *17*, 150–237 (1931)

Nägeli, W.: Der Wind als Standortsfaktor bei Aufforstungen in der subalpinen Stufe. Mitt. Schweiz. Anst. Forstl. Versuchswes. *47*, 35–147 (1971)

Napp-Zinn, K.: Anatomie des Blattes. I. Blattanatomie der Gymnospermen: Handbuch der Pflanzenanatomie. Zimmermann, W., Ozenda, P., Wulff, H.D. (eds.). Vol. 8, Teil 1. Berlin-Nikolassee: Gebr. Borntraeger, 1966

Nather, H.: Zur Keimung der Zirbensamen. Zentralbl. Gesamte Forstwes. *75*, 61–70 (1958)

Negisi, K.: Photosynthesis, respiration and growth in 1-year-old seedlings of *Pinus densiflora, Cryptomeria japonica* and *Chamaecyparis obtusa*. Bull. Tokyo Univ. For. *62*, 1–115 (1966)

Neilson, R.E., Ludlow, M.M., Jarvis, P.G.: Photosynthesis in Sitka spruce (*Picea sitchensis* (BONG.) CARR.). II. Response to temperature. J. Appl. Ecol. *9*, 721–745 (1972)

Neuwinger, I.: Ökologische Untersuchungen in der subalpinen Stufe. Teil I. Bodenfeuchtemessungen. Mitt. Forstl. Bundersversuchsanst. Mariabrunn *59*, 257–264 (1961)

Neuwirth, G.: Gasstoffwechselökologische Provenienzunterschiede bei Fichte (*Picea abies* L.). Arch. Forstwesen *18*, 1325–1339 (1969)

Neuwirth, G., Garelkov, D., Klemm, W., Klemm, M., Naumov, S., Welkov, D.: Ökologisch-physiologische Untersuchungen in Waldbeständen Westbulgariens. Arch. Forstwes. *15*, 379–428 (1966)

Newbould, P.J.: Methods for estimating the primary production of forests. IBP Handbook No 2. Oxford-Edinburgh: Blackwell Sci. Publ., 1967

Noble, D.L., Alexander, R.R.: Environmental factors affecting natural regeneration of Engelmann Spruce in the central Rocky Mountains. For. Sci. *23*, 420–429 (1977)

Oberarzbacher, P.: Beiträge zur physiologischen Analyse des Höhenzuwachses von verschiedenen Fichtenklonen entlang eines Höhenprofils im Wipptal (Tirol) und in Klimakammern. Diss. Univ. Innsbruck, 1977

Oswald, H.: Beobachtungen über die Samenverbreitung bei der Zirbe *(Pinus cembra)*. Allg. Forstz. Wien *67*, 200–202 (1956)

Oswald, H.: Verteilung und Zuwachs der Zirbe *(Pinus cembra* L.) der subalpinen Stufe an einem zentralalpinen Standort. Mitt. Forstl. Bundesversuchsanst. Mariabrunn *60*, 437–499 (1963)

Oswald, H.: Conditions forestières et potentialité de l'épicéa en haute ardèche. Ann. Sci. For. *26*, 183–224 (1969)

Ott, E.: Über die Abhängigkeit des Radialzuwachses und der Oberhöhen bei Fichte und Lärche von der Meereshöhe und Exposition im Lötschertal. Schweiz. Z. Forstwes. *129*, 169–193 (1978)

Ovington, J.D.: Dry matter production of *Pinus sylvestris* L. Ann. Bot. *21*, 287–314 (1957)

Patten, D.T.: Vegetational pattern in relation to environments in the Madison Range, Montana. Ecol. Monogr. *33*, 375–406 (1963)

Patzelt, G., Bortenschlager, S.: Die postglazialen Gletscher- und Klimaschwankungen in der Venedigergruppe (Hohe Tauern, Ostalpen). Z. Geomorphol. N.F. *16*, 25–72 (1973)

Pharis, R.P., Hellmers, H., Schuurmans, E.: Kinetics of the daily rate of photosynthesis at low temperatures for two conifers. Plant Physiol. *42*, 525–531 (1967)

Pisek, A.: Die photosynthetischen Leistungen von Pflanzen besonderer Standorte. Pflanzen der Arktis und des Hochgebirges. In: Handbuch der Pflanzenphysiologie, Band V/2. Ruhland, W. (ed.). Berlin-Göttingen-Heidelberg: Springer, 1960, pp. 376–414

Pisek, A., Cartellieri, E.: Zur Kenntnis des Wasserhaushaltes der Pflanzen. III. Alpine Zwergsträucher. Jahrb. Wiss. Bot. *79*, 131–190 (1933)

Pisek, A., Cartellieri, E.: Zur Kenntnis des Wasserhaushaltes der Pflanzen. IV. Bäume und Sträucher. Jahrb. Wiss. Bot. *88*, 22–68 (1939)

Pisek, A., Kemnitzer, R.: Der Einfluß von Frost auf die Photosynthese der Weißtanne *(Abies alba* Mill.). Flora (Jena) B *157*, 314–326 (1968)

Pisek, A., Knapp, H.: Zur Kenntnis der Respirationsintensität von Blättern verschiedener Blütenpflanzen. Ber. Dtsch. Bot. Ges. *72*, 287–294 (1959)

Pisek, A., Larcher, W.: Zusammenhang zwischen Austrocknungsresistenz und Frosthärte bei Immergrünen. Protoplasma *44*, 30–46 (1954)

Pisek, A., Larcher, W., Moser, W., Pack, I.: Kardinale Temperaturbereiche der Photosynthese und Grenztemperaturen des Lebens der Blätter verschiedener Spermatophyten. III. Temperaturabhängigkeit und optimaler Temperaturbereich der Netto-Photosynthese. Flora (Jena) B *158*, 608–630 (1969)

Pisek, A., Larcher, W., Pack, I., Unterholzner, R.: Kardinale Temperaturbereiche der Photosynthese und Grenztemperaturen des Lebens der Blätter verschiedener Spermatophyten. II. Temperaturmaximum der Netto-Photosynthese und Hitzeresistenz der Blätter. Flora (Jena) B *158*, 110–128 (1968)

Pisek, A., Larcher, W., Unterholzner, R.: Kardinale Temperaturbereiche der Photosynthese und Grenztemperaturen des Lebens der Blätter verschiedener Spermatophyten. I. Temperaturminimum der Nettoassimilation, Gefrier- und Frostschadensbereiche der Blätter. Flora (Jena) B *157*, 239–264 (1967)

Pisek, A., Rehner, G.: Temperaturminima der Netto-Assimilation von mediterranen und nordisch-alpinen Immergrünen. Ber. Dtsch. Bot. Ges. *71*, 188–193 (1958)

Pisek, A., Schiessl, R.: Die Temperaturbeeinflußbarkeit der Frosthärte von Nadelhölzern und Zwergsträuchern an der alpinen Waldgrenze. Ber. Naturwiss. Med. Ver. Innsbruck *47*, 33–52 (1946)

Pisek, A., Sohm, H., Cartellieri, E.: Untersuchungen über osmotischen Wert und Wassergehalt von Pflanzen und Pflanzengesellschaften der alpinen Stufe. Beih. Bot. Zentralbl. *52* B, 634–675 (1935)

Pisek, A., Tranquillini, W.: Transpiration und Wasserhaushalt der Fichte *(Picea excelsa)* bei zunehmender Luft- und Bodentrockenheit. Physiol. Plant. *4*, 1–27 (1951)

Pisek, A., Tranquillini, W.: Assimilation und Kohlenstoffhaushalt in der Krone von Fichten- (*Picea excelsa* LINK) und Rotbuchenbäumen (*Fagus silvatica* L.). Flora *141*, 237–270 (1954)

Pisek, A., Winkler, E.: Assimilationsvermögen und Respiration der Fichte (*Picea excelsa* LINK) in verschiedener Höhenlage und der Zirbe (*Pinus cembra* L.) an der alpinen Waldgrenze. Planta *51*, 518–543 (1958)

Pisek, A., Winkler, E.: Licht- und Temperaturabhängigkeit der CO_2-Assimilation von Fichte (*Picea excelsa* LINK), Zirbe (*Pinus cembra* L.) und Sonnenblume (*Helianthus annuus* L.). Planta *53*, 532–550 (1959)

Platter, W.: Wasserhaushalt, cuticuläres Transpirationsvermögen und Dicke der Cutinschichten einiger Nadelholzarten in verschiedenen Höhenlagen und nach experimenteller Verkürzung der Vegetationsperiode. Diss. Univ. Innsbruck, 1976

Plesník, P.: Horná hranica lesa vo Vysokých a v Belanských Tatrách. Bratislava (1971)

Plesník, P.: Some problems of the timberline in the Rocky Mountains compared with Central Europe. Arct. Alp. Res. *5*, A77–A84 (1973)

Polster, H.: Die physiologischen Grundlagen der Stofferzeugung im Walde. München: Bayer. Landwirtschaftsverlag, 1950

Prutzer, E.: Die Verdunstungsverhältnisse einiger subalpiner Standorte. Mitt. Forstl. Bundesversuchsanst. Mariabrunn *59*, 231–256 (1961)

Pümpel, B.: Über die Auswirkungen von Düngung und Mykorrhizaimpfung auf Wachstum, Entwicklung, Mykorrhizabildung und Frostresistenz junger Fichten und Zirben. Diss. Univ. Innsbruck, 1973

Pümpel, B., Göbl, F., Tranquillini, W.: Wachstum, Mykorrhiza und Frostresistenz von Fichtenjungpflanzen bei Düngung mit verschiedenen Stickstoffgaben. Eur. J. For. Pathol. *5*, 83–97 (1975)

Rehder, H.: Zur Ökologie insbesondere Stickstoffversorgung subalpiner und alpiner Pflanzengesellschaften im Naturschutzgebiet Schachen (Wettersteingebirge). Diss. Bot. 6. Lehre: Cramer, 1970

Rehder, H.: Zum Stickstoffhaushalt alpiner Rasengesellschaften. Ber. Dtsch. Bot. Ges. *84*, 759–767 (1971)

Renvall, A.: Die periodischen Erscheinungen der Reproduktion der Kiefer an der polaren Waldgrenze. Acta For. Fenn. *1* (1912)

Rohmeder, E.: Die Zirbelkiefer (*Pinus cembra*) als Hochgebirgsbaum. Jb. Ver. Schutze Alpenpfl. Tiere (München) *13*, 27–39 (1941)

Rohmeder, E.: Die Bedeutung der Samenherkunft für die Forstwirtschaft im Hochgebirge. In: Forstsamengewinnung und Pflanzenanzucht für das Hochgebirge. Schmidt-Vogt, H. (ed.). München: BLV 1964, pp. 17–35

Rohmeder, E.: Das Saatgut in der Forstwirtschaft. Hamburg, Berlin: Parey, 1972

Röhrig, E., Lüpke, B.: Der Einfluß der ökologischen Bedingungen auf die Keimung von Lärchensamen. Forstarchiv *39*, 194–199 (1968)

Ronco, F.: Influence of high light intensity on survival of planted Engelmann spruce. For. Sci. *16*, 331–339 (1970)

Rook, D.A.: The influence of growing temperature on photosynthesis and respiration of *Pinus radiata* seedlings. N.Z.J. Bot. *7*, 43–55 (1969)

Rottenburg, W.: Die Standardisierung von Frostresistenzuntersuchungen, angewandt an Außenepidermiszellen von *Allium cepa* L. Protoplasma *65*, 37–48 (1968)

Rouschal, E.: Die kühlende Wirkung des Transpirationsstromes in Bäumen. Ber. Dtsch. Bot. Ges. *57*, 53–66 (1939)

Sakai, A.: Temperature fluctuation in wintering trees. Physiol. Plant. *19*, 105–114 (1966)

Sakai, A.: Mechanism of desiccation damage of forest trees in winter. Contrib. Inst. Low Temp. Sci. Ser. B, *15*, 15–35 (1968)

Sakai, A., Otsuka, K.: Freezing resistance of alpine plants. Ecology *51*, 665–671 (1970)

Sakai, A., Weiser, C.J.: Freezing resistance of trees in North America with reference to tree regions. Ecology *54*, 118–126 (1973)

Scharfetter, R.: Beiträge zur Kenntnis subalpiner Pflanzenformationen. Österr. Bot. Z. *67*, 1–14, 63–96 (1918)

Scharfetter, R.: Das Pflanzenleben der Ostalpen. Wien und Leipzig: Deuticke, 1938

Schiechtl, H.M.: Physiognomie der Waldgrenze im Gebirge. Allg. Forstz. (Wien) *77*, 105–111 (1966)

Schmidt, E.: Baumgrenzenstudien am Feldberg im Schwarzwald. Tharandter Forstl. Jahrb. *87*, 1–43 (1936)

Schmidt-Vogt, H.: Die Zapfen- und Samenreifung im Hochgebirge. In: Forstsamengewinnung und Pflanzenanzucht für das Hochgebirge. Schmidt-Vogt, H. (ed.). München: BLV, 1964, pp. 168–177

Schmidt-Vogt, H.: Die Fichte, Band 1. Hamburg und Berlin: Paul Parey, 1977

Schmidt-Vogt, H., Gross, K.: Untersuchungen zum winterlichen Gaswechsel der Fichte (*Picea abies* L. Karst.) unter Freilandbedingungen. Allg. Forst. Jagdz. *147*, 189–192 (1976)

Schröter, C.: Das Pflanzenleben der Alpen. Zürich: Albert Raustein, 1926

Schulze, E.D.: Der CO_2-Gaswechsel der Buche (*Fagus silvatica* L.) in Abhängigkeit von den Klimafaktoren im Freiland. Flora *159*, 177–232 (1970)

Schulze, E.D., Mooney, H.A., Dunn, E.L.: Wintertime photosynthesis of bristlecone pine (*Pinus aristata*) in the White Mountains of California. Ecology *48*, 1044–1047 (1967)

Schwarz, W.: Der Einfluß der Tageslänge auf die Frosthärte, die Hitzeresistenz und das Photosynthesevermögen von Zirben und Alpenrosen. Diss. Univ. Innsbruck, 1968

Schwarz W.: Der Einfluß der Photoperiode auf das Austreiben, die Frosthärte und die Hitzeresistenz von Zirben und Alpenrosen. Flora *159*, 258–285 (1970)

Schwarz, W.: Das Photosynthesevermögen einiger Immergrüner während des Winters und seine Reaktivierungsgeschwindigkeit nach scharfen Frösten. Ber. Dtsch. Bot. Ges. *84*, 585–594 (1971)

Semichatowa, O.A.: Über die Atmung von Hochgebirgspflanzen (Russian). In: Botanische Probleme VII. Moskau-Leningrad: Nauka, 1965, pp. 142–158

Sharpe, D.M.: The effective climate in the dynamics of alpine timberline ecosystems in Colorado. Diss. Southern Illinois Univ., 1968

Shidei, T.: Productivity of Haimatsu (*Pinus pumila*) community growing in alpine zone of Tateyama Range. J. Jpn. For. Soc. *45*, 169–173 (1963)

Sigmond, J.: Kambiumtätigkeit und Spätholzentwicklung bei der Fichte (*Picea excelsa*) an Standorten unterschiedlicher Höhenlage. Beih. Bot. Zentralbl. *54*, Abt. A, 531–568 (1936)

Silen, R.R.: Lethal surface temperatures and their interpretation for Douglas-fir. Diss. Abstr. *21*, 404 (1960)

Slatyer, R.O.: Water deficits in timberline trees in the Snowy Mountains of south-eastern Australia. Oecologia *24*, 357–366 (1976)

Slatyer, R.O.: Altitudinal variation in the photosynthetic characteristics of Snow Gum, *Eucalyptus pauciflora* Sieb. ex Spreng. III. Temperature response of material grown in contrasting thermal environments. Aust. J. Plant Physiol. *4*, 301–312 (1977)

Slatyer, R.O., Morrow, P.A.: Altitudinal variation in the photosynthetic characteristics of Snow Gum, *Eucalyptus pauciflora* Sieb. ex Spreng. I. Seasonal changes under field conditions in the Snowy Mountains Area of South-eastern Australia. Aust. J. Bot. *25*, 1–20 (1977)

Smith, E.M., Hadley, E.B.: Photosynthetic and respiratory acclimation to temperature in *Ledum groenlandicum* populations. Arct. Alp. Res. *6*, 13–27 (1974)

Spomer, G.G., Salisbury, F.B.: Eco-physiology of geum turbinatum and implications concerning alpine environments. Bot. Gaz. *129*, 33–49 (1968)

Štastný, T.: Photoperiodic reaction testing in *Larix decidua* Mill. Acta Inst. For. Zvolenensis 1971, 29–56 (1971)

Steiner, M.: Winterliches Bioklima und Wasserhaushalt der Pflanzen an der alpinen Baumgrenze. Bioklim. Beibl. *2*, 57–65 (1935)

Stern, R.: Der Waldrückgang im Wipptal. Mitt. Forstl. Bundesversuchsanst. Mariabrunn *70* (1966)

Stern, R.: Versuche mit Nadelholz-Saaten auf subalpinen Standorten. Mitt. Forstl. Bundesversuchsanst. Wien *96*, 51–59 (1972)

Stocker, O.: Die Abhängigkeit der Transpiration von den Umweltfaktoren. In: Handbuch der Pflanzenphysiologie, Band III. Ruhland, W. (ed.). Berlin-Göttingen-Heidelberg: Springer, 1956, pp. 436-488

Townsend, A.M., Hanover, J.W., Barnes, B.V.: Altitudinal variation in photosynthesis, growth, and monoterpene composition of western white pine (*Pinus monticola* Dougl.) seedlings. Silvae Genet. *21*, 133–139 (1972)

Tranquillini, W.: Der Ultrarot-Absorptionsschreiber im Dienste ökologischer Messungen des pflanzlichen CO_2-Umsatzes. Ber. Dtsch. Bot. Ges. *65*, 102–112 (1952)

Tranquillini, W.: Über den Einfluß von Übertemperaturen der Blätter bei Dauereinschluß in Küvetten auf die ökologische CO_2-Assimilationsmessung. Ber. Dtsch. Bot. Ges. *67*, 191–204 (1954)

Tranquillini, W.: Die Bedeutung des Lichtes und der Temperatur für die Kohlensäureassimilation von *Pinus cembra* Jungwuchs an einem hochalpinen Standort. Planta *46*, 154–178 (1955)

Tranquillini, W.: Standortsklima, Wasserbilanz und CO_2-Gaswechsel junger Zirben (*Pinus cembra* L.) an der alpine Waldgrenze. Planta *49*, 612–661 (1957)

Tranquillini, W.: Die Frosthärte der Zirbe unter besonderer Berücksichtigung autochthoner und aus Forstgärten stammender Jungpflanzen. Forstwiss. Zentralbl. *77*, 89–105 (1958)

Tranquillini, W.: Die Stoffproduktion der Zirbe (*Pinus cembra* L.) an der Waldgrenze während eines Jahres. Planta *54*, 107–151 (1959)

Tranquillini, W.: Beitrag zur Kausalanalyse des Wettbewerbs ökologisch verschiedener Holzarten. Ber. Dtsch. Bot. Ges. *75*, 353–364 (1962)

Tranquillini, W.: Die Abhängigkeit der Kohlensäureassimilation junger Lärchen, Fichten und Zirben von der Luft- und Bodenfeuchte. Planta *60*, 70–94 (1963a)

Tranquillini, W.: Climate and water relations of plants in the subalpine region. In: The water relations of plants. Rutter, A.J., Whitehead, F.H. (eds.). Oxford: Blackwell Sci. Publ., 1963b, pp. 153–167

Tranquillini, W.: Photosynthesis and dry matter production of trees at high altitudes. In: The formation of wood in forest trees. M.H. Zimmermann (ed.). New York: Academic Press, 1964a, pp. 505–518

Tranquillini, W.: Blattemperatur, Evaporation und Photosynthese bei verschiedener Durchströmung der Assimilationsküvette. Mit einem Beitrag zur Kenntnis der Verdunstung in 2000 m Seehöhe. Ber. Dtsch. Bot. Ges. *77*, 204–218 (1964b)

Tranquillini, W.: Das Klimahaus auf dem Patscherkofel im Rahmen der forstlichen Forschung. In: Industrieller Pflanzenbau, Band II, Vortragsreihe 2. Symposium Industrieller Pflanzenbau Wien, 1965a, pp. 147–154

Tranquillini, W.: Über den Zusammenhang zwischen Entwicklungszustand und Dürreresistenz junger Zirben (*Pinus cembra* L.) im Pflanzengarten. Mitt. Forstl. Bundesversuchsanst. Mariabrunn *66*, 241–271 (1965b)

Tranquillini, W.: Über die physiologischen Ursachen der Wald- und Baumgrenze. Mitt. Forstl. Bundesversuchsanst. Wien *75*, 457–487 (1967)

Tranquillini, W.: CO_2-Gaswechsel und Umwelt. In: Klimaresistenz, Photosynthese und Stoffproduktion. Tagungsberichte Nr. 100 der DAL Berlin, 1968, pp. 153–169

Tranquillini, W.: Photosynthese und Transpiration einiger Holzarten bei verschieden starkem Wind. Zentralbl. Gesamte Forstwes. *86*, 35–48 (1969)

Tranquillini, W.: Der Wasserhaushalt junger Forstpflanzen nach dem Versetzen und seine Beeinflußbarkeit. Zentralbl. Gesamte Forstwes. *90*, 46–52 (1973)

Tranquillini, W.: Der Einfluß von Seehöhe und Länge der Vegetationszeit auf das cuticuläre Transpirationsvermögen von Fichtensämlingen im Winter. Ber. Dtsch. Bot. Ges. *87*, 175–184 (1974)

Tranquillini, W.: Water relations and alpine timberline. In: Water and plant life. Ecol. Studies 19. Lange, O.L., Kappen, L., Schulze, E.D. (eds.). Berlin-Heidelberg-New York: Springer, 1976, pp. 473–491

Tranquillini, W., Holzer, K.: Über das Gefrieren und Auftauen von Coniferennadeln. Ber. Dtsch. Bot. Ges. *71*, 143–156 (1958)

Tranquillini, W., Lechner, F., Oberarzbacher, P., Unterholzner, L., Holzer, K.: Über das Höhenwachstum von Fichtenklonen in verschiedener Seehöhe. Mitt. Forstl. Bundesversuchsanst. Wien, in press (1978)

Tranquillini, W., Machl-Ebner, I.: Über den Einfluß von Wärme auf das Photosynthesevermögen der Zirbe (*Pinus cembra* L.) und der Alpenrose (*Rhododendron ferrugineum* L.) im Winter. Rep. Kevo Subarctic Res. Stat. *8*, 158–166 (1971)

Tranquillini, W., Schütz, W.: Über die Rindenatmung einiger Bäume an der Waldgrenze. Zentralbl. Gesamte Forstwes. *87*, 42–60 (1970)

Tranquillini, W., Turner, H.: Untersuchungen über die Pflanzentemperaturen in der subalpinen Stufe mit besonderer Berücksichtigung der Nadeltemperaturen der Zirbe. Mitt. Forstl. Bundesversuchsanst. Mariabrunn *59*, 127–151 (1961)

Tranquillini, W., Unterholzner, R.: Das Wachstum zweijähriger Lärchen einheitlicher Herkunft in verschiedener Seehöhe. Zentralbl. Gesamte Forstwes. *85*, 43–59 (1968)

Troll, C.: Ökologische Landschaftsforschung und vergleichende Hochgebirgsforschung. Wiesbaden: Steiner, 1966

Troll, C.: The upper timberlines in different climatic zones. Arct. Alp. Res. *5*, A3–A18 (1973)

Tschermak, L.: Waldbau auf pflanzengeographisch-ökologischer Grundlage. Wien: Springer, 1950

Turner, H.: Über das Licht- und Strahlungsklima einer Hanglage der Ötztaler Alpen bei Obergurgl und seine Auswirkung auf das Mikroklima und auf die Vegetation. Arch. Meteorol. Geophys. Bioklimatol. *8* B, 273–325 (1958a)

Turner, H.: Maximaltemperaturen oberflächennaher Bodenschichten an der alpinen Waldgrenze. Wetter und Leben *10*, 1–12 (1958b)

Turner, H.: Jahresgang und biologische Wirkungen der Sonnen- und Himmelsstrahlung an der Waldgrenze der Ötztaler Alpen. Wetter und Leben *13*, 93–113 (1961a)

Turner, H.: Ökologische Untersuchungen in der subalpinen Stufe. 8. Die Niederschlags- und Schneeverhältnisse. Mitt. Forstl. Bundesversuchsanst. Mariabrunn *59*, 265–315 (1961b)

Turner, H.: Über Schneeschliff in den Alpen. Wetter und Leben *20*, 192–200 (1968)

Turner, H.: Grundzüge der Hochgebirgsklimatologie. In: Die Welt der Alpen. Innsbruck: Pinguin—Frankfurt: Umschau, 1970, pp. 170–182

Turner, H.: Mikroklimatographie und ihre Anwendung in der Ökologie der subalpinen Stufe. Ann. Meteorol. N.F. *5*, 275–281 (1971)

Turner, H., Rochat, P., Streule, A.: Thermische Charakteristik von Hauptstandortstypen im Bereich der oberen Waldgrenze (Stillberg, Dischmatal bei Davos). Mitt. Eidg. Anst. Forstl. Versuchswes. *51*, 95–119 (1975)

Turner, H., Tranquillini, W.: Die Strahlungsverhältnisse und ihr Einfluß auf die Photosynthese der Pflanzen. Mitt. Forstl. Bundesversuchsanst. Mariabrunn *59*, 69–103 (1961)

Ulmer, W.: Über den Jahresgang der Frosthärte einiger immergrüner Arten der alpinen Stufe, sowie der Zirbe und Fichte. Jahrb. Wiss. Bot. *84*, 553–592 (1937)

Unterholzner, L.: Höhenzuwachs und Knospenentwicklung bei verschiedenen österreichischen Fichtenherkünften mit besonderer Berücksichtigung der Ausreifungsvorgänge. Diss. Univ. Innsbruck, 1978

Vander Wall, S.B., Balda, R.P.: Coadaptations of the Clark's nutcracker and the pinon pine for efficient seed harvest and dispersal. Ecol. Monogr. *47*, 89–111 (1977)

Vareschi, V.: Die Gehölztypen des obersten Isartales. Ber. Naturwiss. Med. Ver. Innsbruck *42*, 79–184 (1931)

Vincent, G., Vincent, J.: Ergebnisse des internationalen Fichtenprovenienzversuches. Silvae Genet. *13*, 141–146 (1964)

Vorreiter, L.: Bau und Festigkeitseigenschaften des Holzes der Glatzer Schneebergfichte. Thar. Forstl. Jahrb. *88*, 65–126, 235–285, 351–385 (1937)

Walter, H.: Standortslehre. Stuttgart: Ulmer, 1951

Walter, H.: Die Vegetation der Erde in öko-physiologischer Betrachtung. Band II. Jena: VEB Gustav Fischer, 1968

Walter, H.: Ökologische Verhältnisse und Vegetationstypen in der intermontanen Region des westlichen Nordamerikas. Verh. Zool. Bot. Ges. Wien, *110/111*, 111–123 (1971/1972)

Walter, H., Medina, E.: Die Bodentemperatur als ausschlaggebender Faktor für die Gliederung der subalpinen und alpinen Stufe in den Anden Venezuelas. Ber. Dtsch. Bot. Ges. *82*, 275–281 (1969)

Wardle, J.A.: Ecology of *Nothofagus solandri*. N.Z. J. Bot. *8*, 494–646 (1970)

Wardle, P.: A comparison of alpine timberlines in New Zealand and North America. N. Z. J. Bot. *3*, 113–135 (1965)

Wardle, P.: Engelmann spruce (*Picea engelmannii* ENGEL.) at its upper limits on the Front Range, Colorado. Ecology *49*, 483–495 (1968)

Wardle, P.: An explanation for alpine timberline. N. Z. J. Bot. *9*, 371–402 (1971)

Wardle, P.: New Zealand timberlines. Tussock Grasslands and Mountain Lands. Inst. Rev. *23*, 31–48 (1972)

Wardle, P.: Alpine timberlines. In: Arctic and alpine environments. Ives, J.D., Barry, R.G. (ed.). London: Methuen and Co Ltd 1974, pp. 371–402

Waring, R.H., Cleary, B.D.: Plant moisture stress: evaluation by pressure bomb. Science *155*, 1248–1254 (1967)

West, M., Mooney, H.A.: Photosynthetic characteristics of three species of sagebrush as related to their distribution patterns in the White Mountains of California. Am. Midl. Nat. *88*, 479–484 (1972)

Willstätter, R., Stoll, A.: Untersuchungen über die Assimilation der Kohlensäure. Berlin: Springer, 1918

Winkler, E.: Klimaelemente für Innsbruck (582 m) und Patscherkofel (1909 m) im Zusammenhang mit der Assimilation von Fichten in verschiedenen Höhenlagen. Veröff. Museum Ferdinandeum Innsbruck *37*, 19–48 (1957)

Wright, R.D.: Some ecological studies on Bristlecone pine in the White Mountains of California. Diss. Univ. Calif. Los Angeles, 1963

Yoda, K.: Comparative ecological studies on three main types of forest vegetation in Thailand. III. Community respiration. Nat. Life Southeast Asia *5*, 83–148 (1967)

Yoshino, M.M.: Wind-shaped trees in the subalpine zone in Japan. Arct. Alp. Res. *5*, Pt. *2*, A115–A126 (1973)

Żelawski, W., Góral, I.: Seasonal changes in the photosynthesis rate of Scots pine (*Pinus silvestris* L.) seedlings grown from seed of various provenances. Acta Soc. Bot. Pol. *35*, 587–598 (1966)

Żelawski, W., Kucharska, J.: Winter depression of photosynthetic activity in seedlings of Scots pine (*Pinus silvestris* L.). Photosynthetica *1*, 207–213 (1967)

Żelawski, W., Niwiński, Z.: Variability of some needle characteristics in scots pine (*Pinus silvestris* L.) ecotypes grown in native conditions. Ekol. Pol. Ser. A, *14*, 301–308 (1966)

Zimmermann, M.H.: Effect of low temperature on ascent of sap in trees. Plant Physiol. *39*, 568–572 (1964)

Zinn, B.: Abhängigkeit der Lignin- und Xylembildung von äußeren Faktoren. Diss. Univ. Basel, 1930

Žumer, M.: Growth rhythm of some forest trees at different altitudes. (Norwegian) Meld. Nor. Landbrukshøgsk. *48*, 1–31 (1969)

Taxonomic Index

Subject Index

Ecological Studies
Analysis and Synthesis

Editors: W. D. Billings, F. Golley, O. L. Lange, J. S. Olson

A Selection

Springer-Verlag
Berlin
Heidelberg
New York

F. Hallé, R. A. A. Oldeman, P. B. Tomlinson

Tropical Trees and Forests

An Architectural Analysis

1978. 111 figures, 10 tables. XV, 441 pages
ISBN 3-540-08494-0

Recently, the ecological study of the tropical rain forest from the view point of vegetative architecture proved to be a rewarding subject, and continued research led to further results particularly in the field of Sylvignesis. This book provides an up-to-date synthesis, bringing this new and promising field of endeavor within the reach of biologists, ecologists, foresters, and teachers.

From the contents:
Elements of tree architecture: The initiation of the tree, apical meristems and buds. Extension growth in tropical trees. Phyllotaxis and shoot symmetry. Branching. Branch polymorphism. Abscission. Inflorescence. Radial growth. Root systems in tropical tress. – *Inherited tree architecture:* The concept of architecture and architectural tree models. Illustrated key to the architectural models of tropical trees. Descriptions of architectural tree models. Architecture of lianes. Architecture of herbs. Architecture of fossil trees. – *Opportunistic tree architecture:* Reiteration. Energetics. Growth potential of forest trees. A note on floristics. – *Forests and vegetation:* The architecture of forest plots. Sylvigenesis.

Springer-Verlag
Berlin
Heidelberg
New York

F. H. Bormann, G. E. Likens

Pattern and Process in a Forested Ecosystem

Disturbance, Development and the Steady State Based on the Hubbard Brook Ecosystem Study

1979. 74 figures. Approx. 275 pages
ISBN 3-540-90321-6

For the last 15 years the authors have intensively studied the hydrology, biochemistry and ecology of six small watershed-ecosystems at the Hubbard Brook Experimental Forest. Drawing heavily on these well-known studies, they present in this book an integrated picture of the structure, function, and development over time of the northern hardwood ecosystem in Northern New England. Important ecological problems are examined, such as the limits to primary production and biomass accumulation, the relationship between species diversity and stabiliy, variations in biogeochemical behavior over time, and the effect of the weathering-erosion interaction on productivity. The authors present the viewpoint that, at this stage of evolution in ecosystem science, a carefully defined and documented case history provides the best means for determining the principles of ecosystem structure and function, and – most importantly from a holistic standpoint – ecologically sound programs of landscape management can be developed.

Contents:
The Nothern Hardwood Forest: A Model for Ecosystem Development. Energetics, Biomass, Hydrology, and Biogeochemistry of the Aggrading Ecosystem. – *Reorganization:* Loss of Biotic Regulation. – *Development of Vegetation Following Clear Cutting:* Species Strategies and Plant Community Dynamics. – *Reorganization:* Recovery of Biotic Regulation. Ecosystem Development and the Steady State. The Steady Stade as a Component of the Landscape. Forest Harvest and Landscape Management.